C Programming

2nd Edition

by Dan Gookin

C Programming For Dummies®, 2nd Edition

Published by: **John Wiley & Sons, Inc.**, 111 River Street, Hoboken, NJ 07030-5774, www.wiley.com

For general information on our other products and services, please contact our Customer Care Department within the U.S. at 877-762-2974, outside the U.S. at 317-572-3993, or fax 317-572-4002. For technical support, please visit https://hub.wiley.com/community/support/dummies.

Wiley publishes in a variety of print and electronic formats and by print-on-demand. Some material included with standard print versions of this book may not be included in e-books or in print-on-demand. If this book refers to media such as a CD or DVD that is not included in the version you purchased, you may download this material at http://booksupport.wiley.com. For more information about Wiley products, visit www.wiley.com.

Library of Congress Control Number: 2020945155

ISBN: 978-1-119-74024-7; 978-1-119-74025-4 (ebk); 978-1-119-74026-1 (ebk)

Manufactured in the United States of America

SKY10021425_092320

Contents at a Glance

Table of Contents

Introduction

When I was in school, I'd open a new math textbook and look in the back, marveling at the problems. Someday, I thought, I would understand all this nonsense.

You should do that with this book right now: Open it up to one of the final chapters. Look over the C programming code and think to yourself, "Someday soon, this will all make perfect sense to me."

Say "Hello, world" to *C Programming For Dummies*, 2nd Edition, the book that sets you on the path to become a computer programmer. Once despised vermin, banished to basement server rooms and suffering from a lack of personal hygiene, programmers are now valued and contributing members of society. Some are billionaires. And they all started their careers by learning to program.

The C language lets you master a number of electronic gizmos. You can craft your own programs, dictating your every whim and desire to computers, tablets, and cell phones. The electronics dutifully obey. Given the information offered in this book, you can pass that programming class, impress your friends, be admired by Hollywood, or even start your own software company. Truly, learning to program is a worthy investment of your time.

This book helps make learning how to program understandable and enjoyable. You don't need any programming experience — you don't even need to buy new software. You just need the desire to program in C and the ability to have fun while doing so.

Why the C Language?

An argument surfaces every few years that learning C is a road to nowhere. Newer, better programming languages exist, and it's far better to learn them than to waste time learning C.

Poppycock.

C continues to dominate the charts for best and most useful programming languages, often beating out the newer languages the cool programmers use. Further, C is like the Latin of computer languages: Just about every Johnny-come-lately programming language uses C syntax. C keywords and even certain functions find their way into other popular languages, from C++ to Java to Python to whatever the latest, trendy language might be.

My point is that once you learn C programming, all those other programming languages come easy. In fact, many of the books that teach those other languages often assume that you know a little C before you start out. This assumption is frustrating for a beginner — but not when you already know C.

So ignore the lofty pundits and know-it-all poohbahs. C is still relevant. Programming for microcontrollers, operating systems, and major software packages is still done using good ol' C. You are not wasting your time.

The C Programming For Dummies Approach

As a programmer, I've toiled through many programming books. I know what I really don't like to see, and, lamentably, I see it often — that is, where the author writes pages-long code or boasts about what he knows, impressing his fellow nerds and not really teaching anything. Too much of this type of training exists, which is probably why you've picked up this book.

My approach here is simple: Short programs. To-the-point demonstrations. Lots of examples. Plenty of exercises.

The best way to learn something is by doing it. Each concept presented in this book is coupled with sample code. The listings are short enough that you can quickly type them in — and I recommend that you do so. You can then build and run the code to see how things work. This immediate feedback is not only gratifying, it's also a marvelous learning tool.

Sample programs are followed by exercises you can try on your own, testing your skills and expanding your knowledge. Suggested answers to the exercises and all the source code can be found on this book's companion website:

```
https://c-for-dummies.com/cprog
```

How This Book Works

This book teaches the C programming language. It starts out by assuming that you know little to nothing about programming, and it finishes by covering some of the more advanced C operations.

To program in C, you need a computer. This book makes no assumptions about the computer you select: It can be a Windows PC, a Macintosh, or a Linux system. You can choose to use an integrated development environment (IDE) such as Code::Blocks, or you can compile and run the sample programs at a command prompt.

This book also wastes no time, getting you started immediately in Chapter 1. Nothing is introduced without a full explanation first. Due to the nature of programming, I've made a few exceptions; they're carefully noted in the text. Otherwise, the book flows from front to back, which is how best to read this book.

C language keywords and functions are shown in italic text, as in *printf()* and *break*. Some keywords, such as *for* and *if*, may make the sentence read in a goofy way, which is why those words are shown in italic.

Filenames and variable names are shown in `monofont` type, such as `program.exe` and `loop`.

If you need to type something, that text is shown in bold. For example, "Type the **blorfus** command" means that you should type *blorfus* at the keyboard. You're directed when to press the Enter key, if at all.

When working numbered steps, text to type appears in regular (roman) type:

3. **Type** exit **and press the Enter key.**

You type the word *exit* and then press the Enter key.

Program samples are shown as snippets on the page, similar to this one:

```
if( i == 1)
    printf("eye won");
```

You don't need to type an example unless you're directed to do so.

Full program listings are shown and numbered in each chapter; for example:

LISTING 1-1: The Code::Blocks Skeleton

```
#include <stdio.h>
#include <stdlib.h>

int main()
{
    printf("Hello world!\n");
    return(0);
}
```

Because of this book's margins, text in a listing may occasionally wrap, extending from one line to the next. You do not need to split up your code in a similar manner, and I remind you whenever such a thing occurs.

The listings in this book don't contain line numbers, but your text editor might. This book references the sample code listings by using the line numbers, which you can also use in your editor to examine the code.

Exercises are numbered by chapter and then sequentially. So the third exercise in Chapter 13 is Exercise 13-3. You're directed in the text to work an exercise. Here's an example:

Exercise 1-1: Type the source code from Listing 1-1 into your editor. Save it under the filename `ex0101.c`. Build and run.

Answers for all exercises can be found on the web:

```
https://c-for-dummies.com/cprog
```

Go to this web page if you want to copy-and-paste the source code as well.

Icons Used in This Book

REMEMBER

This icon flags information worthy enough to remember. Though I recommend remembering as much as you can, these icons flag the stuff you just can't forget.

TIP

A tip is a suggestion, a special trick, or something super fancy to help you out.

WARNING

This icon marks something you need to avoid. It's advice that could also be flagged with a Tip or Remember icon but has dire consequences if ignored.

TECHNICAL STUFF

Face it: All of programming is technical. I reserve the use of this icon for extra-technical tidbits, asides, and anecdotes. Call it "nerd stuff."

Parting Thoughts

I enjoy programming. It's a hobby, and I find it incredibly relaxing, frustrating, and rewarding. I assume that you share these feelings, though you may also be a struggling student or someone who wants a career. Regardless, *enjoy* programming. If you can imagine the program you want to write on a screen, you can make it happen. It may not happen as fast as you like, but it can happen.

Please work the exercises in this book. Try some on your own, variations on a theme. Continue working at problems until you solve them. The amazing thing about programming is that no single, absolutely correct way to do something exists. Anytime you try, you're learning.

If possible, find a programming friend who can help you. Don't make them do the work or explain how things run, but rely on them as a resource. Programming can be a solo thing, but it's good to occasionally commiserate with others who also program in C — or in any language.

This book has a few companion websites. The primary one is found here:

```
https://c-for-dummies.com/cprog
```

You can also check out my C programming blog, which is updated every Saturday with new lessons and offers a monthly Exercise challenge:

```
https://c-for-dummies.com/blog
```

The publisher also features a companion website, which I'm obliged to mention here, though it's not updated as frequently as my own site. Visit `www.dummies.com` and type **C programming** into the search box to find this book's support page and other goodies.

For more help, or just to say hi, you can send me email at

```
dan@c-for-dummies.com
```

I'm happy to hear from you, though I won't write code for you. I also cannot explain university assignments. (I don't do B-trees. No one does.) And if you have any questions specific to this book — especially any errors or typos — feel free to pass them along.

Enjoy your C programming!

1

The ABs of C

IN THIS PART . . .

Get started with C coding

Work through your very first program

Learn how programming works

Discover the parts of the C language

Craft a basic C skeleton

» Programming at the command prompt

» Using an IDE

» Creating a command prompt program

» Working in Code::Blocks

» Compiling a program

Chapter 1

A Quick Start for the Impatient

You're most likely eager to get started programming in C. I shan't waste your time.

TIP

If you already have a compiler or an IDE set up and are ready to go, skip to Chapter 2.

What You Need to Program

The two most important things you need to begin your programming adventure are

- A computer
- Access to the Internet

The computer is your primary tool for writing and compiling code. Yes, even if you're writing a game for the Xbox, you need a computer to be able to code. The computer can be a PC or a Macintosh. The PC can run Windows or Linux.

Internet access is necessary to obtain the programming software. You need a text editor to write the code and a compiler to translate the code into a program. The compiler generally comes with other tools you need, such as a linker and a debugger. All these tools are found at no cost on the Internet.

The software tools offer two approaches to programming: command line and IDE.

If you want to learn C programming as I did back in the dark ages, you use a terminal window and traditional command-line tools: editor, compiler, and linker. The process is fast, but complicated because you're using text mode commands. Still, it offers a spiritual connection with those who built the foundations upon which the computer industry roosts.

The most common way to craft code, however, is to obtain an integrated development environment — called an *IDE* by the cool kids. It combines all the tools you need for programming into one compact, terrifying, and intimidating unit.

Don't freak! The terms *compiler, linker, debugger*, and *terrifying* are all defined in Chapter 2.

Command Prompt Programming

To re-create the environment where the C language was born, use a Unix or Linux terminal window running a shell program such as *bash*. This environment is available to all major computing platforms, and the programming tools used are reliable and well-documented. Programming at the command prompt earns you a nerd merit badge and the admiration of your peers.

For Windows 10, open the Microsoft Store app and install Ubuntu, a free Linux shell. Ensure that you follow the directions to install the Windows Subsystem for Linux, which is an extra step you'll probably miss.

For Linux, you're ready to go: Open a terminal window to access the shell.

For Mac OS X, use the Terminal app. I also recommend obtaining the Homebrew package manager. Visit `https://brew.sh` for directions. Homebrew allows you to install programming tools not available to OS X.

For an editor, you can use any text mode editor available at the command prompt, such as *vi* or *emacs*. You can also "mix it up" and use a window-based editor. I'm fond of using the Windows version of the VIM editor while I simultaneously work at the command prompt in an Ubuntu terminal window.

A C compiler comes native to a Unix/Linux command prompt. The standard version is *cc* or *gcc*, but I recommend that you use the shell's package manager to acquire the LLVM *clang* compiler. In Ubuntu Linux for Windows 10, type this command to install *clang:*

```
sudo apt-get install clang
```

Type your account password to initiate the process. To verify the installation, type

```
clang --version
```

Various Linux distros offer similar package managers, which you can use to obtain an editor and the *clang* compiler.

- The VIM editor can be obtained from `vim.org`.
- Your choice to use the command prompt means you're taking on an extra layer of complexity when it comes to programming. I find it fast and enjoyable, but if you believe it to be too much, especially when first learning the C language, rely instead upon an IDE, as covered in the next section.

IDE Programming

Plenty of programming IDEs are available for your C coding pleasure. On the Mac, use Xcode, which you can install from the App Store. For Windows and Linux, I recommend obtaining the Code::Blocks IDE, which is found at `codeblocks.org`. You can choose any other IDE you prefer, but Code::Blocks for Windows is fairly stable and comes with everything you need — providing that you install the correct version.

Installing Code::Blocks

The Code::Blocks website will doubtless be altered over time, so follow these general steps to install the IDE and confirm that the C compiler is accessible:

1. **On the main Code::Blocks website page, click the Downloads link.**
2. **Click the binary release link.**

   The "binary release" means you're installing a runnable program, not source code or something equally strange.
3. **Choose the proper installation program for your computer's operating system.**

   For Windows 10, I recommend that you choose the installation with the text *mingw-setup* appended. For example:

   ```
   codeblocks-20.03mingw-setup.exe
   ```

   The `20.03` part of the filename is the release number, which will change in the future. The `mingw-setup` choice means you're downloading both the IDE and the MinGW compiler.

   For Linux, click the link to install the proper version for your distro, but keep in mind that Code::Blocks might be more easily acquired by using the Linux GUI package/software manager.
4. **Open the downloaded archive to extract the Code::Blocks IDE installer.**

   In Windows, you see a User Account Control warning when you open the archive. Click Yes to proceed with installation.
5. **Run the installation program.**

   Perform a default installation; you need not customize anything.
6. **Choose to run Code:Blocks: Click the Yes button.**

   Code::Blocks appears, showing its splash screen. Don't start coding now. Instead, confirm the compiler's installation:
7. **Choose Settings, Compiler.**

   The Compiler Settings dialog box appears.
8. **With Global Compiler Settings chosen on the left, click the Toolchain Executables tab on the right side of the dialog box.**
9. **Ensure that the Compiler's Installation Directory text box is filled.**

   In a default confirmation, the following pathname is listed:

   ```
   C:\Program Files (x86)\CodeBlocks\MinGW
   ```

If the text box is blank, use the Browse button (three dots to the right of the text box) to locate the `MinGW` installation directory.

10. Confirm that gcc.exe **is set in the Compiler text box.**

If not, click the Browse button (three dots) to locate the `gcc.exe` program, installed in the `MinGW\bin` directory by default.

11. Close the Compiler Settings dialog box; click OK.

Installation is complete. I recommend you close the Code::Blocks window. Finish the installation program as well.

Touring the Code::Blocks workspace

Figure 1-1 illustrates the Code::Blocks *workspace*, which is the official name of the massive mosaic of windows and toolbars you see arranged on the screen.

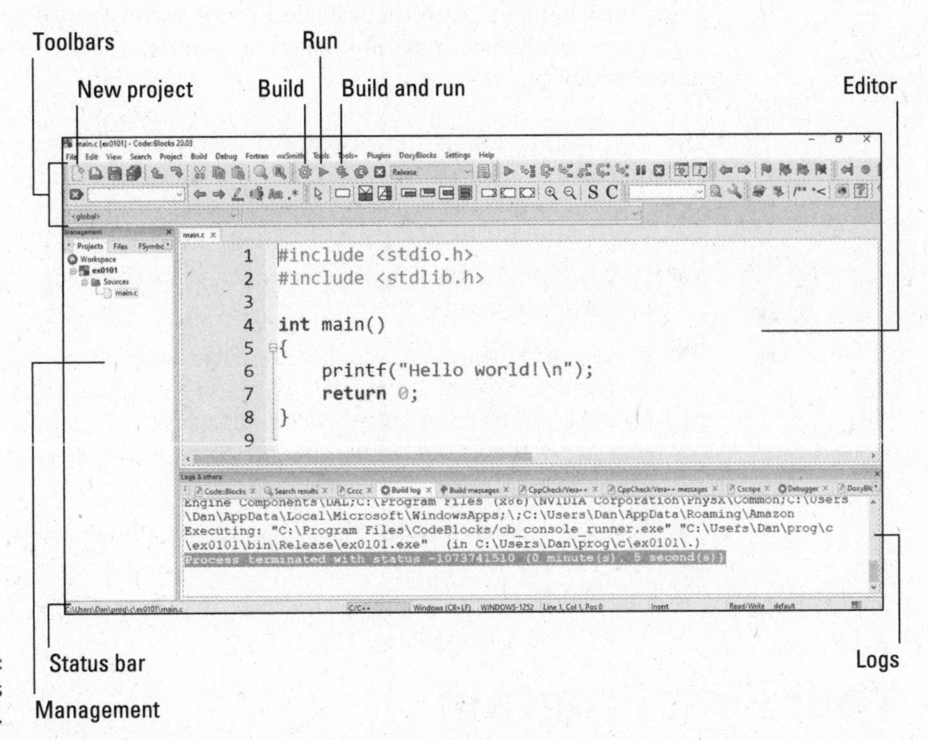

FIGURE 1-1: The Code::Blocks workspace.

On your computer, as well as in Figure 1-1, locate the following parts of the workspace:

Toolbars: These messy strips, adorned with various command buttons, cling to the top of the Code::Blocks window. The toolbars can be rearranged or hidden, so don't mess with them until you get comfy with the interface.

Management: The pane on the left side of the workspace features four tabs, though you may not see all four at one time. The window provides a handy oversight of your programming endeavors.

Status bar: At the bottom of the screen, you see information about the project, editor, and other activities that take place in Code::Blocks.

Editor: The big window in the center-right area of the screen is where you type code.

Logs: The bottom of the screen features a window with many, many tabs displaying various tidbits about the programming process. The tab used most often is named Build Log.

The View menu controls the visibility of every item displayed in the window. Choose the pane name, such as Manager, from the View menu to show or hide that item. Control toolbars by using the View, Toolbars submenu.

TIP

- Maximize the Code::Blocks program window so that it fills the screen. You need all that real estate.
- In addition to color-coding your text, the Code::Blocks editor offers an autocomplete feature. Items that must be typed in pairs, such as quotes, parentheses, and so on, are generated automatically for you. Certain C language keywords and functions are presented automatically, along with hints for their arguments and options.
- The editor features a tabbed interface, which lets you work on multiple source code files at one time.

Your First Program

The traditional first program written for any programming language is called Hello World. It's boring, like all traditions.

Listing 1-1 shows the Hello World source code as presented in Code::Blocks. This code is generated by default whenever you start a new Code::Blocks project.

LISTING 1-1: The Code::Blocks Skeleton

```
#include <stdio.h>
#include <stdlib.h>

int main()
{
    printf("Hello world!\n");
    return 0;
}
```

The programming process works the same whether you use the command prompt or an IDE:

1. **Use an editor to write the source code.**
2. **Compile and link the source code into a program.**
3. **Run the program to see whether it works.**

Chapter 2 goes over these steps in detail, but I assume you're in a rush. The following sections cover the specifics for the command prompt and IDE environments.

Coding at the command prompt

Use your text editor to create a new file and type in the text presented in Listing 1-1. This code, like all source code in this book, is available on the companion website: `https://c-for-dummies.com/cprog`.

Save the source code file as `ex0101.c`. The filename must end in `.c` ("dot C") to be recognized as a C source code file by the compiler.

Compile and link, or "build," the source code into a program file in a Unix terminal window, such as the Ubuntu bash shell in Windows 10. Type the following command in the same folder/directory where the source code file is saved:

```
clang -Wall ex0101.c
```

The name of the compiler program is *clang*. The `-Wall` argument activates all warning messages. The final argument is the name of the source code file, `ex0101.c`, in this example.

Upon success, you see no output. If you see warnings or error messages, you probably mistyped the source code. Try again: re-edit and compile.

To run the program, type the default program filename `a.out`. In a terminal window, the program name must be prefixed by `./` ("dot slash") to direct the command interpreter to look in the current directory:

```
./a.out
```

You see the message `Hello, world!` output in the terminal window. Now skim to Chapter 2 to discover what just happened.

Building a new Code::Blocks project

Two approaches can be used in Code::Blocks to build the sample code shown in this book: You can build a project for each exercise or you can use an empty file to write the code.

Create a project

For a project, follow these steps:

1. **Choose File, New, Project.**
2. **Choose Console Application and then click the Go button.**
3. **Select C as the programming language and then click the Next button.**

   C is quite different from C++ — you can do things in one language that aren't allowed in the other.
4. **Type a project title.**

   The code exercises in this book follow a naming pattern you can use for the project title: **ex** followed by a 2-digit chapter number and 2-digit sequential number. For the sample code presented in Listing 1-1, the project title is **ex0101**.

   When you set the project title, the project's filename is automatically filled in.
5. **Click the . . . (Browse) button to the right of the text box titled Folder to Create Project In.**

   I recommend that you create and use a special folder for all projects in this book. Once the location is set, you can skip this step in the future.

6. **Click the Next button.**

   The next screen (the final one) allows you to select a compiler and choose whether to create Debug or Release versions of your code, or both.

   The compiler selection is fine; the GNU GCC Compiler (or whatever is shown in the window) is the one you want.

7. **Remove the check mark by Create Debug Configuration.**

   You create this configuration only when you need to *debug,* or fix, a programming predicament that puzzles you. See Chapter 25.

8. **Click the Finish button.**

Code::Blocks builds a skeleton of your project, which is the same code shown in Listing 1-1. Skim to the later section "Building and running."

Create an empty file

A less complicated method for working with this book's exercises is to type the source code into an empty project file. Obey these steps in the Code::Blocks IDE:

1. **Choose File, New, Empty File.**
2. **Type the source code into the editor.**

   You can also copy-and-paste the source code from the companion website into the editor. Refer to the introduction for details on the companion website.

TIP

   If you'd prefer to see the editor color-code your text, save the file! Press Ctrl+S and choose the folder where you plan to keep all this book's programming projects. Name the file according to this book's convention: **ex** followed by a 2-digit chapter number and 2-digit project number. End the filename in **.c** ("dot C") to identify it as a C source code file.

3. **Save the source code file.**

   If the file is already saved, press Ctrl+S. Otherwise, follow the naming convention listed under Step 2.

After the empty file is created, you can build and run the program just as you would had you run through the bothersome technique of starting a Code::Blocks project.

TIP

You can also open the source code file directly in Code::Blocks, if you've obtained it from the companion website: Use the Open command to open the file. At this point, you can build and run, modify the code, or do whatever your heart pleases.

Building and running

In Code::Blocks, the process of compiling and linking C language source code is accomplished in one step called Build.

To build the project, click the Build button on the toolbar, as shown in the margin and illustrated in Figure 1-1. Any compiler warnings or errors appear on the Build Log tab at the bottom of the window. For Listing 1-1, no warnings or errors should be generated, providing the code is input exactly.

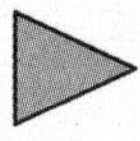

To test-run the program, click the Run button, shown in the margin and illustrated in Figure 1-1. Output appears in a terminal window, as shown in Figure 1-2.

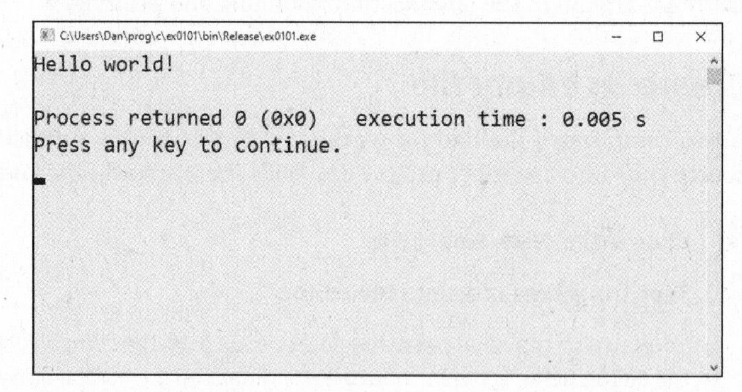

FIGURE 1-2: Program output.

Press the Enter key to close the output window. In Linux, you must type the **exit** command to close the window.

Both build and run steps are combined when you click the Build and Run button, shown in the margin.

When you're done with a project, or even after you've changed a minor thing, save it. Save the source code file, but also save the project if one was created: Choose File, Save Everything to save both the source code and project files.

- » Commands for building and running the code are also found on the Build menu.
- » The keyboard shortcuts for Build, Run, and Build and Run are Ctrl+F9, Ctrl+F10, and F9, respectively. No need to memorize those shortcuts; they're listed on the menu.

TIP

C LIBRARY DOCUMENTATION

You can view documentation for the various C library functions in two ways.

First, in a terminal window, use the *man* program to look up function definitions. For example, type *man printf* to read about the *printf()* function. This documentation is referred to as the "man pages," where "man" is short for *manual.*

The *man* program also documents shell commands. Occasionally, a shell command shares the name of a C language function, such as *stat*. To ensure that you're viewing the proper *man* page, use the *page* argument to specify either page 2 or page 3, where most of the C language functions are found. For example:

man stat — view the *man* page on the *stat* shell command.

man 2 stat — view the *man* page for the C language *stat()* function.

If you're not using a Unix-like shell to program, *man* page documentation is found on the Internet. I recommend these pages:

```
http://man7.org/linux/man-pages/dir_section_2.html
```

```
http://man7.org/linux/man-pages/dir_section_3.html
```

TECHNICAL STUFF

- The program output appears in the top part of the command prompt window (refer to Figure 1-2). The last two lines are generated by the IDE when the program is run. The text shows a value returned from the program to the operating system, a zero, and how long the program took to run (0.005 seconds).

» Reviewing the programming processes

» Creating source code

» Building a program

» Testing the final program

Chapter 2
The Programming Thing

It's called *programming,* though the cool kids know it as *coding* — the process whereby a human being writes information resembling cryptic English that is then somehow translated into directions for an electronic gizmo. In the end, this silent and solitary art grants individuals the power to control electronics. It's a big deal. It's the programming thing.

The History of Programming

Few books written about programming get away with not describing the thrill-a-minute drama of programming history. As a programmer myself, it's difficult not to write about it, let alone contain my enthusiasm at cocktail parties. So consider this section optional reading, though a review of where programming has been and where it is today may help you better understand the art form.

In a nutshell, *programming* is the process of telling a gizmo what to do. That gizmo is *hardware;* the program is *software.*

Reviewing early programming history

The first known machine to be programmed was Charles Babbage's analytical engine, back in 1822. The programming took place by physically changing the values represented by a column of gears. The engine computed the result of some dull, complex mathematical equation.

In the 1940s, early electronic computers were programmed in a similar manner to Babbage's analytical engine. A major difference was that, rather than rearrange physical gears, instructions were hard-wired directly into electric circuitry. "Programming" pretty much meant "rewiring."

Over time, the rewiring job was replaced by rows of switches. Computer instructions were input by throwing switches in a certain way.

Professor John von Neumann pioneered the modern method of computer programming in the 1950s. He introduced decision making into the process, where computers could make if-then choices. Professor von Neumann also developed the concept of the repeating loop and the subroutine.

It was Admiral Grace Hopper who developed the *compiler,* or a program that creates other programs. Her compiler would take words and phrases in English and translate them into computer code. Thus, the programming language was born.

The first significant programming language was FORTRAN, born in the 1950s. Its name came from *for*mula *tran*slator. Other programming languages of the period were COBOL, Algol, Pascal, and BASIC.

REMEMBER

Regardless of the form, whether it's rewiring circuits, flipping switches, or writing a programming language, the result is the same: telling hardware to do something.

Introducing the C language

The C language was developed in 1972 at AT&T Bell Labs by Dennis Ritchie. It combined features from the B and BCPL programming languages but also mixed in a bit of the Pascal language. Mr. Ritchie, along with Brian Kernighan, used C to create the Unix operating system. A C compiler has been part of that operating system ever since.

In the early 1980s, Bjarne Stoustroup used C as the basis of the object-oriented C++ programming language. The ++ (plus-plus) part of the name is kind of an in-joke, which you'll understand better after reading Chapter 11. Mr. Stoustroup intended C++ to be the successor to C. In many ways it is, yet C remains one of the most popular programming languages.

- The B programming language, upon which C is based, was named after the B in Bell Labs.
- BCPL stands for Basic Combined Programming Language.
- The C++ programming language is quite similar to C, but it's not merely an extension or an add-on. It's easier to learn C++ when you know C, but it's not easy to switch between the languages.
- A D programming language exists. Developed in the early 2000s, it's not as popular today as other current languages, such as Python and Java.
- Unfortunately, I have no idea how to pronounce "Bjarne Stoustroup."

The Programming Process

No matter which language you use, certain procedures are common to the programming process. In this manner, learning to program is like learning to cook: You must take things in a certain order, whether the result is crème brûlée or a smoldering pile of egg glop.

Understanding programming

The goal of programming is to create a program. The language is C, and the tools are the editor, compiler, and linker — or an IDE, which combines everything. The result is a program that directs the hardware to do something. That hardware can be a computer, tablet, phone, microcontroller, or whatever.

Step-by-step, the programming process works like this:

1. Write the source code.
2. Compile and link, or build, the source code into a program.
3. Run and test the program.

A human (you) writes source code. The source code is built into a program in two steps: compiling it into object code and then linking the object code with C libraries to build the program. Finally, that program is run.

The reality goes more like this:

1. Write the source code.
2. Compile the source code into object code.
3. Fix warnings and repeat Steps 1 and 2.
4. Link the object code with libraries to build the program.
5. Fix errors and repeat Steps 1 through 4.
6. Run and test the program.
7. Fix bugs by repeating the entire process.

   Or, more frequently, the program runs fine but you want to add a feature or refine an element. Then you repeat everything.

Hopefully, Steps 3, 5, and 7 don't happen often. Still, you do a lot of fixing in the programming cycle.

The good news is that the compiler dutifully reports the errors and even shows you where they are. That's better than tracking down a bug in miles of wires back in the old ENIAC days.

TECHNICAL STUFF

» Despite having "build" as a single step, the compiler still creates object code, and a linker links the object code into a program. If you goof up royally, linking doesn't even happen and you must fix the compiler warnings before taking the next step.

» One of my professional programmer friends said that the art form should be called debugging, not programming.

» Legend has it that the first computer bug was a moth that Grace Hopper found in the wiring of an early computer. There's some doubt about this legend, considering that the word *bug* has been used since Shakespeare's time to describe something quirky or odd.

Writing source code

Source code represents the part of the process that contains the programming language itself. You use a text editor to write a source code file.

In this book, source code is shown in program listings, such as the example in Listing 2-1.

LISTING 2-1: Standard "Hello World" Program

```
#include <stdio.h>

int main()
{
    puts("Greetings, human.");
    return 0;
}
```

Line numbers are not shown in this book's listings. They aren't part of the code, and showing them here can be confusing. The text editor may display line numbers for reference purposes.

In this book, you're directed to type the source code from a listing as part of an exercise. For example:

Exercise 2-1: Start a new project in Code::Blocks named **ex0201**. Or use your text editor to create the source code file `ex0201.c` for compiling and linking at the command prompt.

Do it: Obey Exercise 2-1, either in Code::Blocks, at the command prompt in a terminal window, or in another C language IDE, such as Xcode on the Mac.

1. **Write the source code in the editor, copying it from Listing 2-1.**

   Alternatively, you can copy-and-paste the code from the companion website or use the listing's source code file, downloaded from the website. Refer to this book's introduction for details.

   If you're creating a Code::Blocks project, erase the skeleton that's provided and replace it with the code shown in Listing 2-1.

2. **Save the source code file.**

There. You've just completed the first step in the programming process — writing source code. The next section continues your journey with the compiler.

REMEMBER

- All C source code files end with the `.c` ("dot-see") filename extension.
- In Windows, I recommend that you set the folder display options so that filename extensions appear.
- C++ source code files have the extension .cpp ("dot-see-pee-pee"). I shall refrain from writing a puerile joke here.

» Source code files follow the same naming conventions as any file on a computer. Traditionally, a small program has the same name (but not extension) as the final program. If your program is named `puzzle`, the source code is most likely named `puzzle.c`.

Compiling and linking

Traditionally, the compiler reads text from a source code file and translates that text — a programming language — into something called *object code*. The *linker* then creates the final program. It links the object code file with C language libraries to create a program. All this happens exactly as written — unless a warning or an error crops up along the way.

You can compile and link separately in an IDE or at the command prompt. Fortunately, modern compilers combine both steps in a process called *build*.

To build the project from Listing 2-1 in Code::Blocks, click the Build toolbar button (shown in the margin) or choose Build ⇨ Build from the menu. Any errors present themselves on the Build Log tab as well as on the Build Messages tab.

To build the program at the command prompt, type this line in the same directory (folder) as the source code file (`ex0201.c`):

```
clang -c -Wall ex0201.c
```

First comes the compiler name, *clang*, followed by the `-Wall` switch to report all warnings. The final argument is the source code filename. Upon success, you see no feedback; otherwise, warnings and error messages spill on the screen like fruit from an upturned apple cart.

(Running the program file is covered in the next section.)

» The *object code* file has the same name as the source code file, but ends in `.o` ("dot-o," "little o"). If the source code file is named `ex0201.c` the object code file is named `ex0201.o`. The modern build process deletes this file unless you specifically direct the compiler to generate and retain the object code file.

» As the compiler translates your C code into object code, it checks for common mistakes, missing items, and other issues. If anything is awry, the compiler displays a list of warnings. To fix them, you re-edit the source code and attempt to compile once again.

» The linker brings in the C language library, which is how the final program is built. The libraries contain the actual instructions that tell the computer (or whatever device) what to do. Those instructions are executed based on the shorthand directions found in the object code.

- Any mistakes found by the linker are called *errors.* When errors occur, a program isn't created; errors are fatal.
- Some C programs link in several libraries, depending on what the program does. In addition to the standard C library, you can link libraries for working with graphics, networking, sound, and so on. As you learn more about programming, you'll discover how to choose and link in various libraries. Chapter 24 offers the details.

Running and testing

The next thing to do after building is to run the result. Running is necessary, primarily to demonstrate that the program does what you intend and in the manner you desire.

When the program doesn't work, you must go back and fix the source code and try again. Yes, it's entirely possible to build a program and see no warnings or errors and then find that the thing doesn't work. It happens all the time.

Continue with Exercise 2-1: Run the program.

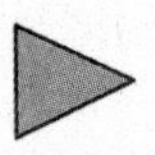

In Code::Blocks, choose Build⇨Run or click the Run toolbar button, shown in the margin.

At the command prompt, type the program name prefixed by the current directory:

```
./a.out
```

PREPROCESSOR DIRECTIVES

Beyond source code, a C compiler also deals with special instructions called *preprocessor directives.* For example, Listing 2-1 shows the following preprocessor directive:

```
#include <stdio.h>
```

The *include* directive instructs the compiler to locate the header file `stdio.h`. The contents of this file are inserted into the source code at compile time. Together, both files are converted into object code.

The name `a.out` is the default C program filename. You can reset this name by specifying the `-o` ("little o") switch followed by the desired output filename when building. For example:

```
clang -Wall ex0201.c -o ex0201
```

The preceding command uses the source code in `ex0201.c` to build a program named `ex0201`. It must be run at the command prompt in the same manner as `a.out`:

```
./ex0201
```

The program runs, outputting text, which appears on the next line after you type the program name:

```
Greetings, human.
```

In Code::Blocks, output appears in a terminal window. Press the Enter key to close the window. For Code::Blocks in Linux, type the **exit** command to close the output terminal window.

A program like `ex0201` doesn't require much testing to ensure that it works. As your programming skills mature, testing, rewriting, compiling, and debugging become a standard part of the process. Don't be discouraged; this is how all programmers learn the craft.

TECHNICAL STUFF

- The mechanics of running a program are carried out by the device's operating system: The operating system loads the program into memory, where the processor or CPU executes the code. This is a loose description of how a program works.
- In Windows, the program name ends with the `exe` (for *exe*cutable) extension, as in `ex0201.exe`. In Mac OS X, Linux, and Unix, the program name has no extension: `ex0201`. Further, in those operating systems, the file's permissions are set for the file to run as an executable.

» Understanding keywords and functions

» Exploring operators, variables, and values

» Learning to comment

» Building a basic C language skeleton

» Printing a math problem

Chapter 3
Anatomy of C

All programming languages consist of instructions that tell a computer or another electronic device what to do. Though basic programming concepts remain the same, each language is different, created to fulfill a specific need or to frustrate a new crop of college freshmen. The C language meets both of those qualifications, being flexible and intimidating. To begin a relationship with C in a friendly, positive way, get to know the language and how it works.

TIP

You'll probably want to reread this chapter after you venture deep into Part 2 of this book.

Parts of the C Language

Unlike a human language, C has no declensions or cases. You'll find no masculine, feminine, or neuter. And you never need to know what the words *pluperfect* and *subjunctive* mean. You do have to understand some of the lingo, the syntax, and other mischief. This section provides an overview of what's what in the C language.

PROGRAMMING LANGUAGE LEVELS

It's almost a tradition. Over time, hundreds of programming languages have been developed. Many fade away, yet new ones pop up annually. The variety is explained by different languages meeting specific needs.

Generally speaking, programming languages exist on three levels:

High-level languages are the easiest to read, using words and phrases found in human languages (mostly English). These languages are quick to learn, but are often limited in their flexibility.

Low-level languages are the most cryptic, often containing few, if any, recognizable human language words. These languages, such as assembly, access hardware directly and therefore are extremely fast. The drawback is that development time is slow because pretty much everything must be done from scratch.

Midlevel languages combine aspects from both high- and low-level languages. As such, these languages are quite versatile, and the programs can be designed to do just about anything. C is the prime example of a midlevel programming language.

Keywords

Forget nouns, verbs, adjectives, and adverbs. The C language has *keywords.* Unlike human languages, where you need to know at least 2,000 or so words to be somewhat literate, the C language sports a scant vocabulary: Only a handful of keywords exist, and you may never use them all. Table 3-1 lists the 44 keywords of the C language.

TABLE 3-1 C Language Keywords

_Alignas	break	float	signed
_Alignof	case	for	sizeof
_Atomic	char	goto	static
_Bool	const	if	struct
_Complex	continue	inline	switch
_Generic	default	int	typedef
_Imaginary	do	long	union

_Noreturn	double	register	unsigned
_Static_assert	else	restrict	void
_Thread_local	enum	return	volatile
auto	extern	short	while

The keywords shown in Table 3-1 represent the C language's basic commands. These simple directions are combined in various interesting ways to do wondrous things. But the language doesn't stop at keywords; continue reading in the next section.

- Don't bother memorizing the list of keywords. Though I still know the 23 "to be" words in English (and in the same order as specified in my eighth grade English text), I've never memorized the C language keywords.
- The keywords are all case-sensitive, as shown in Table 3-1.
- Of the 44 keywords, 32 are original C language keywords. The C99 update (in 1999) added 5 more, and the more recent C11 (2011) update added 7. Most of the newer keywords begin with an underscore, as in *_Alignas*.

WARNING

- Keywords are also known as *reserved words,* which means that you cannot name functions or variables the same as keywords. The compiler moans like a drunken, partisan political blogger when you attempt to do so.

Functions

Where you find only 44 keywords, there are hundreds (if not thousands) of functions in the C language, including functions you write yourself. Think of a *function* as a programming machine that accomplishes a task. Truly, functions are the workhorses of the C language.

The telltale sign of the function is the appearance of parentheses, as in *puts()* for the "put string" function, which displays text. *String* is programming lingo for text that's longer than a single character.

Functions are used in several ways. For example, a *beep()* function may cause a computer's speaker to beep:

```
beep();
```

Some functions are sent values, as in

```
puts("Greetings, human.");
```

Here, the string `Greetings, human.` (including the period) is sent to the *puts()* function, which sends the string to standard output (displayed on the screen). The double quotes define the string; they aren't sent to standard output. The information in the parentheses is said to be the function's *arguments*, or *values*. They are *passed* to the function.

Functions can *generate*, or return, information as well:

```
value = rand();
```

The *rand()* function generates a random number, which is returned from the function and stored in the variable named `value`. Functions in C return only one value at a time. They can also return nothing. The function's documentation explains what the function returns.

Functions can also be sent information and return something:

```
result = sqrt(256);
```

The *sqrt()* function is sent the value 256. It then calculates the square root of that value. The calculation returned is stored in the `result` variable.

WARNING

- See the later section "Variables and values" for a discussion of what a variable is.
- A function in C must be defined before it's used. That definition is called a *prototype*. It's necessary for the compiler to understand whether your code is properly using the function.
- Lists of, and documentation for, all the C language functions is found online, in what are called *C library references*.
- Library function prototypes are held in *header files*, which must be included in your source code. See the later section "Adding a function."
- The mechanics of the functions are stored in C language libraries. A *library* is a collection of functions and the code that executes those functions. When you link your program, the linker incorporates the functions' object code into the final program.
- As with keywords, function names are case-sensitive.

Operators

Mixed in with functions and keywords are various symbols collectively known as *operators.* Most of them are mathematic in origin, including traditional symbols like the plus (+), minus (−), and equal (=) signs.

Operators get thrown in with functions, keywords, and other parts of the C language; for example:

```
result = sqrt(value) + 5;
```

Here, the = and + operators are used to concoct some sort of mathematical mumbo jumbo.

REMEMBER

Not all C language operators perform math. Appendix C lists the lot.

Variables and values

A program works by manipulating information stored in variables. A *variable* is a container into which you can stuff values, characters, or other forms of information. The program can also work on specific, unchanging values that I call *immediate* values:

```
result = sqrt(value) + 5;
```

In this example, `result` and `value` are variables; their content is unknown by looking at the code, and the content can change as the program runs. The number 5 is an immediate value.

C sports different types of variables, each designed to hold specific values or data types. Chapter 6 explains variables and values in more detail.

Statements and structure

As with human languages, programming languages feature *syntax* — it's the method by which the pieces fit together. Unlike English, where syntax can be determined by rolling dice, the method by which C puts together keywords, functions, operators, variables, and values is quite strict.

The core of the C language is the *statement,* which is similar to a sentence in English. A statement is an action, a direction that the program gives to the hardware. All statements end with a semicolon, the C language equivalent of a period:

```
beep();
```

Here, the single function *beep()* is a statement. It can be that simple. In fact, a single semicolon on a line can be a statement:

```
;
```

The preceding statement does nothing.

Statements in C are executed one after the other, beginning at the top of the source code and working down to the bottom. Ways exist to change that order as the program runs, which are presented elsewhere in this book.

The paragraph-level syntax for the C language involves blocks. These are defined by enclosing statements in a pair of curly brackets, or *braces*:

```
{
    if( money < 0 ) getjob();
    party();
    sleep(24);
}
```

These three statements form a block, held within curly brackets, indicating that they belong together. They're either part of a function or part of a loop or something similar. Regardless, they all go together and are executed one after the other.

You'll notice that the statements held within a block are indented one tab stop. That's a tradition in C, but it's not required. The term *whitespace* is used to refer to tabs, empty lines, and other empty parts of the source code.

Generally, the C compiler ignores whitespace, looking instead for semicolons and curly brackets. For example, you can edit the source code from Listing 2-1 to read:

```
#include <stdio.h>
int main(){puts("Greetings, human.");return 0;}
```

That's two lines of source code where before you saw several. The *include* directive must be on a line by itself, but the C code can be all scrunched up with no whitespace. The code compiles successfully.

Thankfully, most programmers use whitespace to make their code more readable.

REMEMBER

- A common mistake made by beginning C language programmers is forgetting to place the semicolon after a statement. It's also a common mistake made by experienced programmers!
- The compiler is the tool that finds missing semicolons. That's because when you forget the semicolon, the compiler assumes that two statements are really one statement. The effect is that the compiler becomes confused and, therefore, in a fit of panic, flags those lines of source code with a warning or error.

Comments

Some items in your C language source code are parts of neither the language nor the structure. Those are *comments*, which can be information about the program, notes to yourself, or filthy limericks.

Traditional C comments begin with the /* characters and end with the */ characters. All text between these two markers is ignored by the compiler, shunned by the linker, and avoided in the final program.

Listing 3-1 shows an update to the code from project ex0201, where comments have been liberally added.

LISTING 3-1: Overly Commented Source Code

```
/* Author: Dan Gookin */
/* This program displays text on the screen */

#include <stdio.h>     /* Required for puts() */

int main()
{
    puts("Greetings, human.");  /* Displays text */
    return 0;
}
```

Comments in Listing 3-1 appear on a line by itself or at the end of a line. The first two lines can be combined for a multiline comment, as shown in Listing 3-2.

LISTING 3-2: Multiline Comments

```
/* Author: Dan Gookin
   This program displays text on the screen */

#include <stdio.h>     /* Required for puts() */

int main()
{
    puts("Greetings, human.");  /* Displays text */
    return 0;
}
```

All text between the /* and the */ is ignored. Some source code editors, such as the one in Code::Blocks, display commented text in a unique color, which further confirms how the compiler sees and ignores the comment text. If you like, type the source code from Listing 3-2 into a text editor to see how comments work and whether they're highlighted.

A second comment style uses the double-slash (//) characters, which originated in the C++ language but is also available in C. This type of comment affects text on one line, from the // characters to the end of the line, as shown in Listing 3-3.

LISTING 3-3: Double-Slash Comments

```
#include <stdio.h>

int main()
{
    puts("Greetings, human.");  // Displays text
    return 0;
}
```

Don't worry about putting comments in your text at this point, unless you're at a university somewhere and the professor is ridiculously strict about it. Comments are for you, the programmer, to help you understand your code and remember what your intentions are. They come in handy down the road, when you're looking at your code and not fully understanding what you were doing. This situation happens frequently.

Behold the Typical C Program

All C programs feature a basic structure, which is easily shown by looking at the C source code skeleton that Code::Blocks uses to start a new project, as shown in Listing 3-4.

LISTING 3-4: Code::Blocks C Skeleton

```
#include <stdio.h>
#include <stdlib.h>

int main()
{
    printf("Hello world!\n");
    return 0;
}
```

This listing isn't the bare minimum, but it gives a rough idea of the basic C program.

REMEMBER

- Just as you read text on a page, C source code flows from the top down. The program starts execution at the first line, and then the next line, and so on until the end of the source code file. Exceptions to this order include decision-making structures and loops, but mostly the code runs from the top down.
- Decision-making structures are covered in Chapter 8; loops are introduced in Chapter 9.

Understanding C program structure

To better understand how C programs come into being, you can create the simplest, most useless type of C program.

Exercise 3-1: Follow the steps in this section to create a new source code file, named **ex0301.c**. You can use Code::Blocks to create a new project or use a text editor and build the project in a terminal window at the command prompt.

Here are the specific steps:

1. **Edit the source code to match Listing 3-5.**

   That's not a misprint. The source code is blank, empty. If you're using Code::Blocks, erase the skeleton provided.

2. **Save the source code.**
3. **Build.**

   Nothing happens or you see a linker error message.

LISTING 3-5: A Simple Program That Does Nothing

```
```

Because the source code is empty, no object code is generated. Even if a program is created, it's empty and does nothing. That's what you told the compiler to do, and the resulting program did it well.

Setting the main() function

All C programs have a *main()* function. It's the first function that's run when a program starts. As a function, it requires parentheses for its arguments but also curly brackets to hold the function's statements in a block, as shown in Listing 3-6.

Continue with Exercise 3-1: Rebuild the source code `ex0301.c`, as shown in Listing 3-6. Save the project. Build and run.

LISTING 3-6: The *main()* Function

```
main() {}
```

You may see a warning because the *main()* function lacks a data type. That's okay. Otherwise, the program generates no output — which is great! You didn't direct the code to do anything, and it did it well. What you see is the minimum C program. It's also known as the *dummy* program.

» *Main* isn't a keyword; it's a function. It's the required first function in all C language source code.

» Unlike other functions, *main()* doesn't need to be prototyped. It does, however, use specific arguments, which is a topic covered in Chapter 15.

Returning something to the operating system

Proper protocol requires that when a program quits, it provides a value to the operating system. Call it a token of respect. That value is an integer (a whole number), usually zero, but sometimes other values are used, depending on what the program does and what the operating system expects.

Continue with Exercise 3-1: Update the source code for `ex0301.c` to reflect the changes shown in Listing 3-7.

LISTING 3-7: Adding the *return* Statement

```
int main()
{
    return(1);
}
```

First, the *main()* function is declared to be an integer function. The *int* keyword tells the compiler that the function returns, or *generates*, an integer value.

The *return* statement sends the value `1` back to the operating system, effectively ending the *main()* function and, therefore, the program.

Continue with Exercise 3-1: Save, build, and run the project.

The results are similar to the previous run, but if you're using the Code::Blocks IDE, you see the return value of `1` specified in the summary text:

```
Process returned 1 (0x1)
```

If you like, edit the code again and change the return value to something else — say, 5. That value appears in the Code::Blocks output when you run the project.

- Traditionally, a return value of `0` is used to indicate that a program has completed its job successfully.
- Return values of `1` or greater often indicate some type of error, or perhaps they indicate the results of an operation.
- The keyword *return* can be used in a statement with or without parentheses. Here it is without them:

```
return 1;
```

  In Listing 3-7, *return* is used with parentheses. The result is the same. I prefer to code *return* with parentheses, which is how it's shown throughout this book.

Adding a function

C programs should do something. Though you can use keywords and operators to have a program do marvelous things, the way to make these things useful is output. Continue working on this chapter's example:

Continue with Exercise 3-1: Modify the project's source code one final time to match Listing 3-8.

LISTING 3-8: More Updates for the Project

```
#include <stdio.h>

int main()
{
    printf("4 times 5 is %d\n",4*5);
    return(0);
}
```

You're adding three lines. First, add the *include* line, which brings in the *printf()* function's prototype. Second, type a blank line to separate the processor directive from the *main()* function. Third, add the line with the *printf()* function. All functions must be declared before use, and the `stdio.h` file contains the declaration for *printf()*.

Before proceeding, please note these two important items in your source code:

» Ensure that you type the *include* directive exactly as written:

```
#include <stdio.h>
```

This directive tells the precompiler to fetch the header file, `stdio.h`. The header file is required in order to use the *printf()* function.

» Ensure that you type the *printf()* statement exactly as written:

The *printf()* function sends formatted text to the standard output device: the display. It also contains a math problem, `4*5`. The result of that equation is calculated by the computer and then displayed in the formatted text:

```
printf("4 times 5 is %d\n",4*5);
```

You'll find lots of important items in the *printf()* statement, each of which is required: quotes, comma, and semicolon. Don't forget anything!

Later chapters cover the *printf()* function in more detail, so don't worry if you're not taking it all in at this point.

Finally, I've changed the return value from `1` to `0`, the traditional value that's passed back to the operating system.

Continue with Exercise 3-1: Save the project's source code. Build and run.

If you get an error, double-check the source code. Otherwise, the result appears in the terminal window, looking something like this:

```
4 times 5 is 20
```

The basic C program is what you've seen presented in this section, as built upon over the past several sections. The functions you use will change, and you'll learn how things work and become more comfortable as you explore the C language.

WHERE ARE THE FILES?

A C programming project needs more than just the source code: It includes header files and libraries. The header files are called in by using the *include* directives; libraries are brought in by the linker. Don't worry about these files, because the compiler and linker handle the details for you.

Because C comes from a Unix background, traditional locations for the header and library files are used. Header files are found in the `/usr/include` directory (folder). The library files dwell in the `/usr/lib` directory. Those are system folders, so look but *don't touch* the contents. I frequently peruse header files to look for hints or information that may not be obvious from the C language documentation. (Header files are plain text; library files are data.)

If you're using a Unix-like operating system, you can visit those directories and peruse the multitude of files located there. On a Windows system, the files are kept with the compiler; usually, in `include` and `lib` folders relative to the compiler's location.

2

C Programming 101

IN THIS PART . . .

Fix compiler and linker errors

Work with values in a program

Explore variables and storage

Grab input and create output

Control program flow and make decisions

Repeat chunks of code with looping statements

Build your own functions

» Disabling statements by using comments

» Fixing compiler errors

» Using the *printf()* function

» Enjoying escape sequences

» Fixing linking errors

Chapter 4
Trials and Errors

One of the nifty things about computer programming is that feedback is immediate. You're informed right away when something goes awry, by the compiler, the linker, or the program not running the way you intended. Believe it or not, that's the way everyone programs! Errors happen, and even the most experienced programmer expects them.

Display Stuff on the Screen

The best way to get started programming is to create tiny code samples that merely toss up some text on the screen. It's quick. And you learn something while you do it.

Displaying a humorous message

A computer is known as a serious device, so why not add some levity?

Exercise 4-1: Carefully type the code from Listing 4-1 into the editor. Build and run.

LISTING 4-1: Another Humorous Example

```
#include <stdio.h>

int main()
{
    puts("Don't bother me now. I'm busy.");
    return(0);
}
```

If you encounter any warnings or errors, fix them. They could be typos or missing items. Everything you see in the editor must look exactly like the code shown in Listing 4-1.

Upon success, the program runs. Its output looks like this:

```
Don't bother me now. I'm busy.
```

Try to contain your laughter.

Exercise 4-2: Modify the source code from Listing 4-1 so that the message says, "I love displaying text!"

REMEMBER

Solutions for all exercises can be found on the web:

```
www.c-for-dummies.com/begc4d/exercises
```

Introducing the *puts()* function

The *puts()* function streams a string of text to the standard output device.

What the heck does that mean?

For now, consider that the *puts()* function displays text on the screen on a line by itself. Here's the format:

```
#include <stdio.h>

int puts(const char *s);
```

Because that official format looks confusing this early in the book, I offer this unofficial format:

```
puts("text");
```

The *text* part is a string of text — basically, anything sandwiched between the double quotes. It can also be a variable, a concept you don't have to worry about until you reach Chapter 7.

The *puts()* function requires that the source code include the `stdio.h` header file. This header file contains the function's prototype. Header files are added to the source code by using the *include* directive, as just shown and in various examples throughout this chapter.

» The C language handles text in streams, which is probably different from the way you think computers normally handle text. Chapter 13 discusses this concept at length.

» The standard output device is usually the computer's display. Output can be redirected at the operating system level; for example, to a file or another device, such as a printer. This reason is why the technical definition of the *puts()* function refers to standard output and not to the display.

Adding more text

When you need to display another line of text, conjure up another *puts()* function in your source code, as shown in Listing 4-2.

LISTING 4-2: Displaying Two Lines of Text

```
#include <stdio.h>

int main()
{
    puts("Hickory, dickory, dock,");
    puts("The mouse ran up the clock.");
    return(0);
}
```

The second *puts()* function does the same thing as the first. Also, because the first *puts()* function requires the `stdio.h` header file, there's no need to include this line again; one reference does the job for any function that requires the same header file.

Exercise 4-3: Type the source code from Listing 4-2 into the editor. Save, compile, and run.

The output appears on two lines:

```
Hickory, dickory, dock,
The mouse ran up the clock.
```

As long as you use the *puts()* function and enclose the text in double quotes, the resulting program spits out that text, displaying it on the screen. Well, okay, *puts()* sends text to the standard output device. (Feel better, university sophomores?)

REMEMBER

» Include the proper header file to prototype functions. The *puts()* function requires the `stdio.h` header.

» The *include* precompiler directive thrusts the named header file into your source code. It's formatted like this:

```
#include <file.h>
```

In this line, *file* represents the name of the header file. All header files sport the `.h` extension, which must be specified with the header filename in the angle brackets.

» There's no need to include the same header file more than once in a source code file.

Exercise 4-4: Modify Listing 4-2 so that the entire nursery rhyme is displayed.

Here's the full text:

Hickory, dickory, dock,

The mouse ran up the clock.

The clock struck one,

The mouse ran down,

Hickory, dickory, dock.

Yeah, it doesn't really rhyme, so for a bonus, change the fourth line of the output so that it does rhyme!

Commenting out a statement

Comments are used to not only add information, remarks, and descriptions to your source code but also disable statements, as shown in Listing 4-3.

LISTING 4-3: Disabling a Statement

```
#include <stdio.h>

int main()
{
    puts("The secret password is:");
/*  puts("Spatula."); */
    return(0);
}
```

Exercise 4-5: Type the source code shown in Listing 4-3. Type `/*` at the start of Line 6, and then press the Tab key to indent the statement to the same tab stop as on the preceding line. Press Tab at the end of Line 6 before adding the final comment marker: `*/`. Save. Build. Run.

Only the first *puts()* function at Line 5 executes, displaying the following text:

```
The secret password is:
```

The second *puts()* function at Line 6 has been "commented out" and therefore doesn't compile.

Exercise 4-6: Uncomment the second *puts()* statement from your solution to Exercise 4-5. Run the program to see the results.

Exercise 4-7: Comment out the first *puts()* function using the // commenting characters. Build and run again.

Goofing up on purpose

If you haven't yet made a mistake typing source code, it's about time to do so.

Exercise 4-8: Carefully type the source code shown in Listing 4-4. If you've been paying attention, you can probably spot the errors. (***Hint:*** Look at the fifth line.) Don't fix them — not yet.

LISTING 4-4: This Program Goes BOOM

```
#include <stdio.h>

int main()
{
    puts("This program goes BOOM!)
    return(0);
}
```

I'll admit that using a smart editor, such as the one in Code::Blocks, makes it difficult to type the missing double quote. Still, make your source code look like Listing 4-4. I'm trying to show you how errors look so that you can fix them in the future.

Build the program. You see warnings and errors along these lines:

```
ex0408.c: In function 'main':
ex0408.c:5:10: warning: missing terminating " character
     puts("This program goes BOOM!)
          ^
ex0408.c:5:10: error: missing terminating " character
     puts("This program goes BOOM!)
          ^~~~~~~~~~~~~~~~~~~~~~~~~
ex0408.c:6:5: error: expected expression before 'return'
     return(0);
     ^~~~~~
ex0408.c:7:1: error: expected ';' before '}' token
 }
 ^
```

Errors and warnings are referenced by line number in the source code. They may appear several times, depending on the severity and how offended the compiler feels.

Numbers in the following list refer to lines in Listing 4-4 as well as the output just shown:

Line 5: Warning: missing terminating " character

The first warning is caused by the missing double-quote character. It's a warning because the compiler could be mistaken.

Line 5: Error: missing terminating " character

No, the compiler isn't mistaken: the double quote *is* missing. It qualifies as a full-on error, which is stronger than a warning.

Line 6: Error: expected expression before 'return'

This error is still related to Line 5, but it's caught at Line 6. Again, it's the missing double quote, which was expected before the `return` statement.

Line 7: Error: expected ';' before '}' token

The final error is again caused by Line 5, the missing semicolon. This error isn't caught until Line 7, the closing curly bracket.

A program doesn't compile when errors are present; you must fix the problem. Warnings can happen and the code compiles — a program might even be created. Whether the program runs properly, however, is dubious.

- A *warning* is something suspicious that's spotted by the compiler, but it may also be something that works as you intended.
- An *error* is a fatal flaw in the program. When an error occurs, the compiler doesn't create object code or the linker fails to build the program. Either way, you're required to fix the problem.

REMEMBER

- Line numbers referenced in warnings and error messages are an approximation. Sometimes, the issue is on the line that's indicated, but it might also exist earlier in the code.

TECHNICAL STUFF

- It's possible to adjust how sensitive the compiler is regarding warnings. In fact, modern compilers have dozens of options designed to turn on or off various compiling conditions. I recommend activating all warnings, which is the default in Code::Blocks and is achieved at the command prompt by using the `-Wall` switch.

Exercise 4-9: Fix the code from Exercise 4-8 by adding the missing double quote. Compile to see the difference in the error messages.

Exercise 4-10: Fix the code again, adding the missing semicolon.

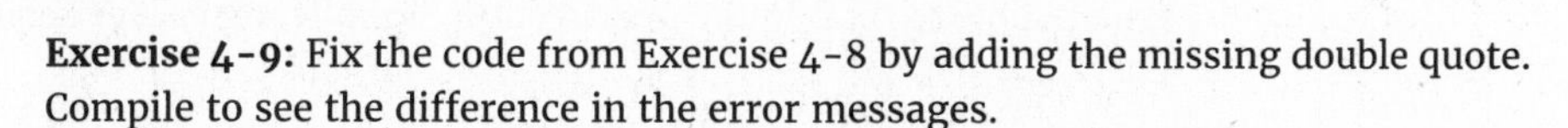

More Text Output Nonsense

The *puts()* function is but one of many functions that sends text to the standard output device. A second, more popular and versatile function is *printf()*. It too displays information to the standard output device, but with more bells and whistles.

Displaying text with *printf()*

On the surface, the *printf()* function looks and works a lot like *puts()*, displaying text to the screen. But *printf()* is far more potent and capable, and you'll probably use it as the primary text output function in your C code, as shown in Listing 4-5.

LISTING 4-5: Using *printf* to Display Text

```
#include <stdio.h>

int main()
{
    printf("I have been a stranger in a strange land.");
    return(0);
}
```

Exercise 4-11: Eagerly type the source code shown in Listing 4-5. Check your typing carefully because you're using a new function, *printf()*, to display text. Save. Build. Run.

The output should look familiar and expected, although there's one tiny difference. If you can spot it, you get a *For Dummies* bonus point. (Don't worry about fixing the problem yet.) If you can't spot the difference, just proceed with Exercise 4-12.

Exercise 4-12: Edit your solution for Exercise 4-4, updating the source code to use the *printf()* function instead of *puts()*.

Don't worry if the output for your solution to Exercise 4-12 doesn't look right. I explain how to fix it in the later section "Employing escape sequences."

Introducing the *printf()* function

The *printf()* function streams a formatted string of text to the standard output device. The official format is a bit overwhelming:

```
#include <stdio.h>

int printf(const char *restrict format, ...);
```

Don't let your eyeballs pop out of your head. Instead, consider my abbreviated format, which pretty much describes how *printf()* is used in this chapter:

```
printf("text");
```

In this definition, *text* is a string of text wedged between double quotes.

The *printf()* function requires the inclusion of the `stdio.h` header file.

The name *printf()* means *print formatted*, and the function really shows its horsepower in displaying formatted output. You can see this feature demonstrated in Chapter 13. The *print* part of the name hails back to the days when C programs sent their output primarily to printers, not to video displays.

Understanding the newline

Unlike the *puts()* function, the *printf()* function doesn't tack a newline character at the end of its output. A *newline* is the character that ends a line of text and directs the terminal to display any following text on the next line — the "new" line.

The following *puts()* function outputs the text `Goodbye, cruel world` on a line by itself:

```
puts("Goodbye, cruel world");
```

The following *printf()* function displays the text `Goodbye, cruel world`:

```
printf("Goodbye, cruel world");
```

After displaying the text, the cursor waits at the space after the `d` in `world`. Any additional text that's displayed appears on the same line, which is what you see for the output of Exercise 4-12:

```
Hickory, dickory, dock,The mouse ran up the clock.The clock
struck one,The mouse ran down,Hickory, dickory, dock.
```

The code does exactly what you programmed it to do, albeit without fully knowing how *printf()* works. The results most likely aren't what you intended.

To make the *printf()* function display text on a line by itself, add the newline character to the end of the text string. Don't bother looking for the newline character on the keyboard; no, it's not the Enter key. You must use a C language escape sequence to type the newline character. Keep reading in the next section.

Employing escape sequences

To reference certain characters that you can't type into your source code, the C language uses something called an escape sequence. The *escape sequence* allows you to direct the compiler to temporarily suspend its acceptance of what you're typing and instead interpret special characters and codes.

The standard escape sequence uses the backslash character followed by a second character; for example:

```
\n
```

This is the escape sequence for the newline character. The compiler reads both the backslash and the character that follows it as a single character, interpreting that character as one that you can't type at the keyboard, such as the Tab key or Enter key or characters that may foul up the source code, such as a double quote.

Table 4-1 lists the standard C language escape sequences.

TABLE 4-1: Escape Sequences

Escape Sequence	Character It Produces
\a	Bell ("beep!")
\b	Backspace, non-erasing
\f	Form feed or clear the screen
\n	Newline
\r	Carriage return
\t	Tab
\v	Vertical tab
\\	Backslash character
\?	Question mark
\'	Single quote
\"	Double quote
\x*nn*	Hexadecimal character code *nn*
\o*nn*	Octal character code *nn*
nn	Octal character code *nn*

Exercise 4-13: Re-edit the source code for your solution to Exercise 4-12, adding the newline character at the end of every *printf()* text string.

An escape sequence is required only when you need a character in a text string and you cannot otherwise type it. For example, if you want to use the statement

```
printf("What!");
```

you don't have to escape the exclamation point character because it doesn't otherwise mess up the text. You would, however, escape a newline, tab, or double-quote character.

Exercise 4-14: Create source code that uses the *printf()* function to display the following line of text:

```
"Hey," said the snail, "I said no salt!"
```

Exercise 4-15: Modify the source code from Exercise 4-14 so that the *puts()* function is used instead of *printf()* to display the same text.

Goofing up on purpose again

The section "Goofing up on purpose," earlier in this chapter, introduces you to compiler errors. The compiler isn't the only part of the program creation process. The other major part is linking, and, yes, you'll find that the linker can detect errors as well, as shown in Listing 4-6.

LISTING 4-6: Another Horrible Mistake

```
#include <stdio.h>

int main()
{
    writeln("Another horrible mistake.");
    return(0);
}
```

Exercise 4-16: Type the source code from Listing 4-6 into the editor. Save. Build. And . . .

The compiler generates a warning for Line 5:

```
implicit declaration of function 'writeln' is invalid in C99
```

It's only a warning because the `writeln()` function hasn't been prototyped — that is, it wasn't found in the `stdio.h` header file. The compiler still generates object code and passes the code to the linker.

The linker, on the other hand, is quite displeased. Here are the linker errors I see on my screen (showing only the relevant portions of the text):

```
/tmp/ex0416-0ea1a6.o: In function `main':
ex0416.c:(.text+0x1c): undefined reference to `writeln'
clang: error: linker command failed with exit code 1 (use -v to
   see invocation)
```

It's the linker's job to bring in a C language library and, specifically, to link in the code for the *writeln()* function. But there is no *writeln()* function, not in the standard C library. Therefore, the program isn't created, and an "undefined reference" error is reported.

- To fix the code, change *writeln()* to *puts()*.
- This type of error occurs most frequently when you define your own functions. This topic is covered in Chapter 10.

» Using *printf()* to display values

» Formatting floating-point output

» Letting the program do the math

» Dealing with integer and float math

Chapter 5

Values and Simple Math

Back in the old days, most people thought of computers in terms of math. Computers calculated rocket trajectories, conducted the census, and screwed up phone bills. They were scary, technological things, and the people who programmed computers were downright geniuses.

Ha! Fooled everyone.

Programmers merely write the equation and punch in a few numbers, and then the computer does the math. That's the genius part. Punching in the wrong numbers is the nongenius part. Before you can get there, you have to understand a bit about values and variables and how the C programming language deals with them.

A Venue for Various Values

Computers deal with both numbers and text. Text comes in the form of individual characters or a parade of characters all tied together in a string. Numbers are pretty much numbers until you get into huge values and fractions. The computer understands everything, as long you properly inform your program of which values are which.

Understanding values

You've probably dealt with numbers all your life, virtually tortured by them throughout your schooling. You may recall terms such as *whole number*, *fraction*, *real number*, and *imaginary number*. Ignore them! These terms mean nothing in computer programming.

When it comes to C programming, you use only two types of numbers:

- Integer
- Float

An *integer* is a whole number — no fractional part. It can be positive. It can be negative. It can be zero, a single digit, or a humongous value such as the amount of money the US government spends in a week (no cents). All these numbers are integers in computer programming jargon.

A *float* is a number that has a fractional part — a decimal place. It can be a very, very small number, like the width of a proton. It can be a very, very large number, like the hat size of the planet Jupiter.

TIP

- Examples of integers: –13, 0, 4, and 234792.
- In programming, you don't type commas in large values.
- Examples of floats are 3.14, 0.1, and 6.023e23. That last number is written in scientific notation, which means that it's the value 6.023×10^{23} — a huge number. (It's *Avogadro's number,* which is another term you've probably forgotten from school.)
- Integers and floats can be either positive or negative.
- Integers are judged by their size, as are floats. The size comes into play when you create storage places for numbers in your programs. Chapter 6 covers the details.
- The term *float* is short for *floating point*. It refers to the method that's used to store large numbers and fractions in the binary counting system of modern electronics.

TO FLOAT OR NOT TO FLOAT

Though it may seem logical to use all floating-point numbers *(floats)* in your programs, the problem is that they're imprecise. In fact, floating-point values are defined by their *precision,* or the number of digits in the number that are truly accurate.

For example, a floating-point value with single-precision accuracy may show from six to nine significant digits in the number. The rest of the numbers in the value are nonsense. That seems sloppy, but for very large or small numbers, it's good enough. When it's not good enough, double-precision accuracy can be used, though such calculations require more processor power.

To put it another way, the value of π as represented using seven significant digits, or single precision, would be accurate enough to define a circle the size of Saturn's orbit, accurate to the millimeter.

Displaying values with *printf()*

The *printf()* function, introduced in Chapter 4, is ideal for displaying not only strings of text but also values. To make that happen, you use *conversion characters* in the function's formatting string. Rather than bore you with a description, consider Exercise 5-1.

Exercise 5-1: Input the source code illustrated in Listing 5-1. Save the file. Build it. Run it.

LISTING 5-1: Displaying Various Values

```
#include <stdio.h>

int main()
{
    printf("The value %d is an integer.\n",986);
    printf("The value %f is a float.\n",98.6);
    return(0);
}
```

The output from Exercise 5-1 looks something like this:

```
The value 986 is an integer.
The value 98.600000 is a float.
```

You're probably surprised that the output doesn't look like this:

```
The value %d is an integer.
The value %f is a float.
```

It doesn't, because the text included in a *printf()* function isn't merely text — it's a formatting string.

The *printf()* function's formatting string can contain plain text, escape sequences, and conversion characters, such as the `%d` in Line 5 and the `%f` in Line 6. These conversion characters act as placeholders for values and variables (arguments) that follow the formatting string.

For the `%d` placeholder, the integer value `986` is substituted. The `%d` conversion character represents decimal integer values.

For the `%f` placeholder, the float value `98.6` is substituted. The `%f` conversion character represents floating-point values. Of course, `98.6` isn't displayed. Instead, you see `98.600000`. This issue is addressed in the later section "Minding the extra zeros."

The `%d` and `%f` are only two of many placeholders for the *printf()* function's formatting string. The rest are covered in Chapter 7.

Exercise 5-2: Create a project that uses the appropriate conversion characters, either `%d` or `%f`, in a *printf()* function to display the following values:

127

3.1415926535

122013

0.00008

REMEMBER

Do not type a comma when specifying a value in your C language source code.

When typing a small floating-point value, remember to prefix the decimal point with a zero, as just shown, with `0.00008`. Likewise, when typing a float value without a decimal part, type the decimal and a zero anyway:

1000000.0

Minding the extra zeros

When you wrote the code for Exercise 5-1, you probably expected the program's output to display the value 98.6, just as it's written. The problem is that you directed the *printf()* function to output that number in the default manner, 98.600000. In fact, you may see more or fewer zeros, depending on your compiler.

The value 98.600000 is a floating-point number, and it most likely represents the way the value is stored inside the computer. Specifically, the value is stored using eight digits, but human beings don't usually write trailing zeros after numbers. Computers? They write as many zeros as fills eight digits (not counting the decimal).

To fix the output, direct the *printf()* function to format the floating-point number. That requires a more complex version of the %f placeholder, something you're introduced to in Chapter 7. Specifically, change the %f placeholder to read %2.1f. Here's the new Line 6 from Listing 5-1:

```
    printf("The value %2.1f is an float.\n",98.6);
```

By squeezing 2.1 between the % and the f, you direct *printf()* to format the output with two digits to the left of the decimal and one digit to the right.

Exercise 5-3: Modify your source code from Exercise 5-2 so that the value 3.1415926535 is displayed by using the %1.2f placeholder, and the value 0.00008 is displayed by using the %1.1f placeholder.

The Computer Does the Math

It should come as no surprise that computers can do math. In fact, I'd bet that your computer is, right now, more eager to solve some mathematical puzzles than it is for you to visit Facebook. Some math examples shown earlier in this chapter merely bored the processor. Time to put it to work!

Doing simple arithmetic

Math in your C source code is brought to you by the +, −, *, and / operators. These are the basic math symbols, with the exception of * and /, mostly because the × and ÷ characters aren't found on the typical computer keyboard. For reference, Table 5-1 lists the basic C language math operators.

TABLE 5-1: Basic Math Operators

Operator	Function
+	Addition
-	Subtraction
*	Multiplication
/	Division

More C math operators exist, as well as a tumult of mathematical functions. Chapter 11 helps you continue exploring math programming potential. For now, the basics will do.

Calculations in C are made by placing values on either side of a math operator, just as you did all throughout school, but with the benefit of the computer making the calculations. Listing 5-2 provides a sample.

LISTING 5-2: The Computer Does the Math

```
#include <stdio.h>

int main()
{
    puts("Values 8 and 2:");
    printf("Addition is %d\n",8+2);
    printf("Subtraction is %d\n",8-2);
    printf("Multiplication is %d\n",8*2);
    printf("Division is %d\n",8/2);
    return(0);
}
```

Exercise 5-4: Input the source code shown in Listing 5-2. Save. Build. Run.

The output looks something like this:

```
Values 8 and 2:
Addition is 10
Subtraction is 6
Multiplication is 16
Division is 4
```

What you see in this code are immediate calculations. That is, the value that's calculated, the *result*, isn't stored. Instead, the program does the math and deals with the result, which is stuffed into the `%d` conversion character in the *printf()* function's formatting text.

Exercise 5-5: Create a program that displays the result of adding 456.98 and 213.4.

Exercise 5-6: Create a program that displays the result of multiplying the values 8, 14, and 25.

Exercise 5-7: Create a program that solves one of those stupid riddles on Facebook: What's the result of 0+50*1−60−60*0+10? Solve the equation yourself before you run the program to see the computer's result. Refer to Chapter 11 to read why the results might be different.

Reviewing the float-integer thing

The difference between an immediate value being a *float* or an *int* is how you specify it in a program. Consider Listing 5-3.

LISTING 5-3: Another Immediate Math Problem

```
#include <stdio.h>

int main()
{
    printf("The total is %d\n",16+17);
    return(0);
}
```

The values 16 and 17 are integers; they have no decimal part.

Exercise 5-8: Create, build, and run the program created from the source code in Listing 5-3.

Building the project yields the answer, which is also an integer:

```
The total is 33
```

Exercise 5-9: Modify the source code to specify one of the values as a float. For example, change Line 5 to read:

```
printf("The total is %d\n",16.0+17);
```

Adding that point-zero doesn't change the value. Instead, it changes the way the number is stored. It's now a *float*.

Save the change in your source code. Build and run.

You may see a warning displayed, depending on whether your compiler is configured. The warning explains that the `%d` placeholder is used to display a floating-point value. You may even see a suggestion to use `%f` instead of `%d`. Regardless, run the program. Here's the result I see:

```
The total is -90871768
```

You may see another value. Regardless, the displayed result is incorrect. That's because the `%d` integer placeholder was used when the calculation includes a *float*. Adding a *float* into any calculation causes the compiler to express the result as a floating-point number. Change Line 5 again, specifying the `%f` placeholder this way:

```
printf("The total is %f\n",16.0+17);
```

Build and run. The result now looks something like this:

```
The total is 33.000000
```

This answer is correct.

Exercise 5-10: Rewrite the source code for Listing 5-3 so that all immediate values are floating-point numbers. Ensure that the *printf()* function displays them with a single digit after the decimal point.

REMEMBER

Anytime a floating-point number is used in a calculation, the result is a floating-point number. Various tricks can be employed to avoid this issue, but for now consider it solid.

Pretending integers are floats

What happens when you attempt to do math with integer values but the result isn't an integer? Don't even think about it! Just admire the code in Listing 5-4:

LISTING 5-4: Where Integers Dare

```
#include <stdio.h>

int main()
{
    printf("%d/%d=%d\n",2,5,2/5);
    return(0);
}
```

Three %d (integer) placeholders are used in the *print()* statement. Each corresponds to an integer value argument in the function: 2, 5, and then the expression 2/5. The placeholders in the format string must match the argument count and the type of value each argument represents, as illustrated in Figure 5-1.

Exercise 5-11: Input the source code from Listing 5-4 into the editor. Build and run to see the crazy result.

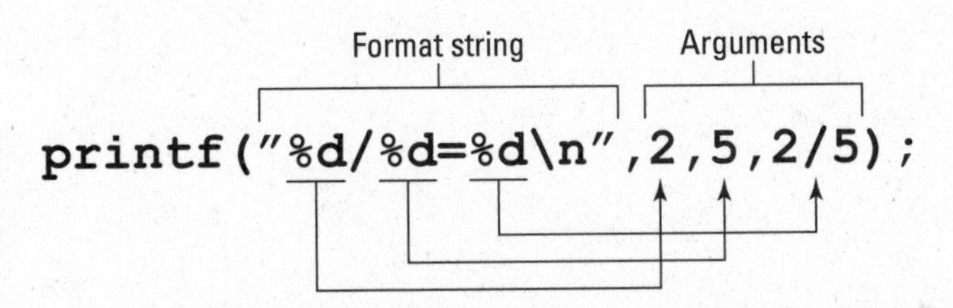

FIGURE 5-1: Matching *printf()* conversion characters and arguments.

Just to be fair, on my calculator the result of 2/5 is equal to 0.4, which is a *float* value. But the computer believes that dividing integer 2 by integer 5 is equal to zero. How would you fix this problem?

Exercise 5-12: Desperately modify the source code from Listing 5-4 so that the %f placeholder is used instead of the final %d in the format string. See if that works. Save, build, and run.

Did you see a warning? If so, the compiler is bemoaning the use of the %f placeholder to output an integer value. Still, a program is created. When I run it, I see:

```
2/5=0.000000
```

Zero again.

The problem here isn't the same as shown earlier in this chapter, when adding a float and an integer value. Instead, the issue is math itself: When you must divide two integer values, unless you're lucky, the result is a float — a value with a decimal portion. The solution is to ensure that the values used are float — even when they're whole numbers.

Exercise 5-13: Rewrite the source code from Listing 5-4 so that the final calculation is made using immediate float values, not integers.

Later in your C journey, you'll learn a trick to typecast calculations, which is a more versatile way to write code that must specify integer values as floats. This topic is covered in Chapter 16.

» Creating a specific data type

» Declaring variables in your code

» Using signed or unsigned integers

» Whipping up multiple variables at a time

» Declaring and assigning at the same time

» Putting variables to work

» Creating constants

Chapter 6
A Place to Put Stuff

Human beings have obsessed over storing stuff ever since the Garden of Eden, when Adam stashed a grape in his belly button. This raises the question of why Adam would have a belly button, but that's not my point. My point is that people enjoy storing things and creating places — boxes, closets, garages, and underground bunkers — in which to store that stuff.

Your C programs can also store things — specifically, various types of information. Computer storage is used to keep these items, but the containers themselves are called *variables*. They're basic components of all computer programming.

Values That Vary

C programs use three types of values: immediate, constant, and variable. An *immediate* value is one that you specify in the source code — a value you type. Constants are discussed later in this chapter. *Variables* are storage containers for values where the contents can change. Their contents can vary — hence the name.

Setting up a quick example

Who likes to read a lot about something before they try it? Not me!

Exercise 6-1: Type the source code shown in Listing 6-1. It uses a single variable, *x*, which is one of the first computer variable names mentioned in the Bible.

LISTING 6-1: Your First Variable

```
#include <stdio.h>

int main()
{
    int x;

    x = 5;
    printf("The value of variable x is %d.\n",x);
    return(0);
}
```

Here's the breakdown of what's going on in Listing 6-1:

Line 5 contains the variable's declaration:

```
int x;
```

All C language variables are declared as a specific data type and assigned a name. The variable in Listing 6-1 is declared as an integer *(int)* data type and given the name *x*.

Line 7 assigns the value 5 to variable *x*:

```
x = 5;
```

The value goes on the right side of the equal sign. The variable goes on the left.

Line 8 uses the variable's value in the *printf()* statement:

```
printf("The value of variable x is %d.\n",x)
```

The %d conversion character is used because the variable's data type is integer. The conversion character must match the data type.

Build and run the code. The output looks like this:

```
The value of variable x is 5.
```

The following sections describe in further detail the mechanics of creating and using variables.

Introducing data types

C language variables are designed to hold specific types of data. If C were a biological programming language, cats and dogs would go into the *animal* data type, and trees and ferns would go into the *plant* data type. C language variables work along these lines, with specific values assigned to matching types of data.

The common C language data types are listed in Table 6-1.

TABLE 6-1 **Basic C Language Variable Types**

Data Type	Description
char	Single-character variable; stores one character of information
int	Integer variable; stores integer (whole number) values
float	Floating-point variable; stores real numbers
double	Floating-point variable; stores very large or very small real numbers
void	No data type

When you need to store an integer value, you use a variable of the integer *(int)* data type. Likewise, if you're storing a letter of the alphabet, you use a variable of the character *(char)* data type. The program's needs are what determine which data types are required for its variables.

- The *char* and *int* data types store integer values. The *char* data type has a shorter range, used primarily to store characters — letters of the alphabet, numbers, and symbols — but it can also be used to store small integer values.
- The *float* and *double* data types are both floating-point variables that can store a tiny or huge value or any value with a decimal portion.

- A newer data type, *_Bool*, stores binary values, 1 or 0, often referred to as TRUE and FALSE, respectively. *_Bool*, a loaner word from C++, must be written with the initial underscore character and a capital *B*. You may not find *_Bool* used in many C program source code listings — most likely, to keep the code compatible with older compilers.

Using variables

Most, if not all, of your future C language programs will employ variables. Earlier in this chapter, Listing 6-1 illustrates the basic three steps for using variables in the C language:

1. Declare the variable, giving it a data type and a name.
2. Initialize the variable (assign it a value).
3. Use the variable.

All three steps are required for working with variables in your code, and these steps must be completed in that order.

To declare a variable, place a statement near the start of a function, such as the *main()* function in every C program. Place the declaration after the initial curly bracket. (Refer to Listing 6-1.) The declaration is a statement on a line by itself, ending with a semicolon:

```
type name;
```

type is the data type: *char*, *int*, and others are introduced throughout this chapter.

name is the variable's name. The name must not be the name of a C language keyword or any other variable name that was previously declared. The name is case-sensitive — though, traditionally, C language variable names are written in lowercase. If you want to be saucy, you can add numbers or underscores to the variable name, but always start a variable name with a letter.

The equal operator assigns a value to a variable. The format is very specific:

```
variable = value;
```

Read this expression as, "The value of `variable` equals `value`."

variable is the variable's name. It must be declared earlier in the source code.

value is either an immediate value, a constant, an expression, another variable, or a value returned from a function. After the statement is executed, the *variable* holds the *value* that's specified.

Assigning a value to a variable satisfies the second step in using a variable. The third step is to do something with the variable. Variables can be used anywhere in your source code that a value could otherwise be specified directly.

In Listing 6-2, four variable types are declared, assigned values, and used in separate *printf()* statements.

LISTING 6-2: Working with Variables

```
#include <stdio.h>

int main()
{
    char c;
    int i;
    float f;
    double d;

    c = 'a';
    i = 1;
    f = 19.0;
    d = 20000.009;

    printf("%c\n",c);
    printf("%d\n",i);
    printf("%f\n",f);
    printf("%f\n",d);
    return(0);
}
```

Exercise 6-2: Type the source code for Listing 6-2 into the editor. Build and run.

The output looks something like this:

```
a
1
19.000000
20000.009000
```

In Line 10, the single-character value a is placed into char variable *a*. Single characters are expressed using single quotes in C: 'a'

In Line 15, you see the %c placeholder used in the *printf()* statement. This placeholder is designed for single characters.

Exercise 6-3: Replace Lines 15 through 18 with a single *printf()* statement:

```
printf("%c\n%d\n%f\n%f\n",c,i,f,d);
```

Build and run the code.

The *printf()* formatting string can contain as many conversion characters as needed, but only as long as you specify the proper quantity and type of variables for those placeholders, and in the proper order. The variables appear after the formatting string, each separated by a comma, as just shown.

Exercise 6-4: Edit Line 12 from your solution to Exercise 6-3 so that the value assigned to variable f is 19.8 and not 19.0. Save this change, build, and run the code.

Did you see the value 19.799999 displayed for variable f? Would you say that the value is imprecise?

Exactly!

The *float* data type is *single precision:* The computer accurately stores only eight digits of the value. The internal representation of 19.8 is really the value 19.799999 because a single-precision *(float)* value is accurate to only the eighth digit. For mathematical purposes, 19.799999 is effectively 19.8; you can direct the code to display this value by using the %.1f placeholder.

Exercise 6-5: Write source code that declares an integer variable blorf and assigns it the value 22. Have a *printf()* statement display the variable's value. Have a second *printf()* statement display this value plus 16. Then have a third *printf()* statement that displays the value of blorf multiplied by itself.

Here's the output from my solution for Exercise 6-5:

```
The value of blorf is 22.
The value of blorf plus 16 is 38.
The value of blorf times itself is 484.
```

Your code's output need not look identical to what's shown here.

TECHNICAL STUFF

» Variables need not be declared at the start of a function. Some programmers declare variables on the line before they're first used. This strategy works, but it's not traditional and can lead to confusion. Most C programmers expect to find all variable declarations at the start of the function.

» It's possible to start a variable name with an underscore, which the compiler believes to be a letter. Even so, variable names that begin with underscores are used internally in the C language. I recommend avoiding this naming convention.

Variable Madness!

I hope that you're getting the hang of the variable thing. If not, please review the first part of this chapter. The variable is truly the heart of any programming language, by allowing you to code flexibility into your programs and have it do amazing things.

Using more-specific data types

The C language offers more data types than are shown in Table 6-1. Depending on the information stored, you may want to use one of these more detailed variable declarations. Table 6-2 lists a buffet of C language data types and the range of values those types can store.

The *value range* column specifies the size of the number you can store in a variable as well as whether negative numbers are allowed. The compiler may not always flag warnings that happen when you assign the wrong value to a variable type. So get it right when you declare the variable!

For example, if you need to store the value `-10`, you use a *short int*, *int*, or *long* variable. You cannot use an *unsigned int*, as the source code in Listing 6-3 demonstrates.

TABLE 6-2

More C Language Data Types

Type	Value Range	*printf()* Conversion Character
_Bool	0 to 1	%d
char	-128 to 127	%c
unsigned char	0 to 255	%u
short int	-32,768 to 32,767	%d
unsigned short int	0 to 65,535	%u
int	-2,147,483,648 to 2,147,483,647	%d
unsigned int	0 to 4,294,967,295	%u
long int	-2,147,483,648 to 2,147,483,647	%ld
unsigned long int	0 to 4,294,967,295	%lu
float	1.17×10^{-38} to 3.40×10^{38}	%f
double	2.22×10^{-308} to 1.79×10^{308}	%f

LISTING 6-3: **Oh, No — an Unsigned *int!***

```
#include <stdio.h>

int main()
{
    unsigned int ono;

    ono = -10;
    printf("The value of ono is %u.\n",ono);
    return(0);
}
```

Exercise 6-6: Create a project with the source code shown in Listing 6-3. Note that the %u conversion character is used for *unsigned int* values. Build and run.

Here's the output:

```
The value of ono is 4294967286.
```

The moral of the story: If your integer variable stores negative numbers, you can't use an *unsigned* variable type.

- The range of the *int* may be the same as the range of the *short int* on some compilers. When in doubt, use a *long*.
- Some compilers support a *long long* integer type, also called "double long." Its values range from –9,223,372,036,854,775,808 to 9,223,372,036,854,775,807 signed, and from 0 to 18,446,744,073,709,551,615 unsigned.
- Specifying the *int* keyword isn't necessary when declaring a *short*, *unsigned*, or *long* integer value.
- The keyword *signed* can be used before any of the integer variable types, as in *signed short int*, though only *short* is necessary.
- The *void* data type is used primarily to declare functions that return no values. Still, it's a valid variable type, found most often when dealing with pointers and memory buffers. See Chapter 10 for information on *void* functions; pointers are avoided until Part 4 of this book.

Working with several variables

I can find nothing in the rules to prevent starting a section with an exercise, so here you go:

Exercise 6-7: Create a program that uses the three integer variables `shadrach`, `meshach`, and `abednego`. Assign integer values to each one, and display the result.

Here's a copy of the output from the program generated by Exercise 6-7. It's my version of the project:

```
Shadrach is 701
Meshach is 709
Abednego is 719
```

Your code can generate different text, but the underlying project should work. And give yourself a bonus if your answer matched my answer, which is given in Listing 6-4.

LISTING 6-4: The Answer to Exercise 6-7

```
#include <stdio.h>

int main()
{
    int shadrach, meshach, abednego;

    shadrach = 701;
    meshach = 709;
    abednego = 719;
    printf("Shadrach is %d\nMeshach is %d\nAbednego is %d\n",shadrach,
    meshach,abednego);
    return(0);
}
```

When declaring multiple variables of the same type, you can specify all of them on the same line, as shown in Listing 6-4 (on Line 5). You don't even have to put spaces after each name; the line could have easily been written

```
int shadrach,meshach,abednego;
```

The C compiler doesn't care about spaces — specifically, whitespace — outside of something enclosed in double quotes.

I also stacked up the results in a single, long *printf()* statement. The line wraps in Listing 6-4 because of this book's page width, and it may wrap in your editor as well. When the code wraps, don't press the Enter key to start a new line.

You can split a long statement in C simply by escaping the Enter key press at the end of a line. *Escaping* in this context doesn't mean that you're fleeing danger (other than offending the compiler); instead, you use the backslash (the escape character) to read the Enter key without messing up the statement in your code. To wit:

```
printf("Shad is %d\nMesh is %d\nAbed is d\n",\
   shadrach,meshach,abednego);
```

I shortened the names in the formatting string so that the text fits on a line on this page. Between *printf()*'s formatting string and the argument list, right after the first comma, I typed a backslash and then pressed the Enter key. The effect is that the line is broken visually, but the compiler still sees it as a single statement.

Assigning a value upon creation

In Listing 6-5, the integer variable `start` is created and assigned the value `0` upon creation. This combination saves typing another line of code that would assign `0` to the `start` variable.

LISTING 6-5: Declaring a Variable and Assigning a Value

```
#include <stdio.h>

int main()
{
    int start = 0;

    printf("The starting value is %d.\n",start);
    return(0);
}
```

Exercise 6-8: Build a program by using the source code shown in Listing 6-5.

Exercise 6-9: Modify the source code for Exercise 6-7 so that the three variables are assigned their values on the same lines where the variables are declared. Two solutions may be possible for this exercise.

Reusing variables

Variables vary, so their contents can be changed at any time in the program. The examples shown elsewhere in this chapter use a variable only once and don't alter its value. That's pretty much the same as a constant (covered later in this chapter), which makes them good examples for learning but a poor representation of reality.

In your programming journey, variables are declared, and then their values may be, well, whatever. Not only that; it's possible to reuse variables over and over — no harm done. That's an easy example to show, as illustrated in Listing 6-6.

LISTING 6-6: Variables Recycled

```
#include <stdio.h>

int main()
{
    int prime;

    prime = 701;
    printf("Shadrach is %d\n",prime);
    prime = 709;
    printf("Meshach is %d\n",prime);
    prime = 719;
    printf("Abednego is %d\n",prime);
    return(0);
}
```

Exercise 6-10: Create new source code from Listing 6-6. As you can see, the variable `prime` is used over and over, each time changing its value. The new value that's assigned replaces any existing value. Build and run the project.

The output from Exercise 6-10 is the same as from Exercise 6-9.

Listing 6-7 illustrates how variables can interact with each other.

LISTING 6-7: Variables Mix It Up

```
#include <stdio.h>

int main()
{
    int a,b,c;

    a = 5;
    b = 7;
    c = a + b;
    printf("Variable c=%d\n",c);
    return(0);
}
```

Line 9 is the one to notice:

```
c = a + b;
```

The value of variable c is assigned the sum of variables a and b. This calculation is made when the program runs, and then the result — whatever weirdo value that could be — is displayed.

Exercise 6-11: Create a project using the source code in Listing 6-7. Can you guess the output?

Exercise 6-12: Create a new source code file using Listing 6-7 as a starting point. Declare three *float* variables and assign values to two of them. Assign a value to the third variable by dividing the first variable by the second variable. Output the result.

Constants Always the Same

Computers and their electronic brethren enjoy doing repetitive tasks. In fact, anything you do on a computer that requires you to do something over and over demands that a faster, simpler solution be at hand. Often, it's your mission to simply find the right tool to accomplish that goal.

Using the same value over and over

It may be too early in your C programming career to truly ponder a repetitive program; the topic of looping is covered in Chapter 9. But that doesn't mean you can't code programs that use values over and over.

Exercise 6-13: Create a new program using the source code shown in Listing 6-8. Save it, build it, run it.

LISTING 6-8: It's a Magic Number

```
#include <stdio.h>

int main()
{
    printf("The value is %d\n",3);
    printf("And %d is the value\n",3);
    printf("It's not %d\n",3+1);
    printf("And it's not %d\n",3-1);
    printf("No, the value is %d\n",3);
    return(0);
}
```

The code uses the value 3 on every line. Here's the output:

```
The value is 3
And 3 is the value
It's not 4
And it's not 2
No, the value is 3
```

Exercise 6-14: Edit the code to replace the value 3 with 5. Compile and run.

You might think that Exercise 6-14 is cruel and requires a lot of work (unless your source code editor has a nifty search-and-replace command), but such changes occur frequently in programming. For example, I wrote a program that displays the top three most recent items added to a database. But then I wanted to change the list so that it shows the top five items. As you had to do in Exercise 6-14, I had to painstakingly search-and-replace throughout the entire source code, carefully plucking out specific references to 3 and substituting 5.

There must be a better way.

Constants in your code

One solution for dealing with the problems presented in Exercises 6-13 and 6-14 is to use a variable. However, because the variable's value doesn't change throughout the *main()* function, you can instead use a constant.

Like a variable, a *constant* is a named value used throughout a function. Think of it as a substitute, something you can quickly change in one spot and have the effect widespread.

Constants are created like variables, but with the keyword *const* prefixing the data type:

```
const type name = value;
```

type is the data type, the same as a variable. It can be *int*, *char*, *double*, or any valid C language data type; refer to Table 6-2.

name is the constant's name, just like a variable with the same naming restrictions, rules, and whatnot.

The *value* must be assigned to the constant as it's declared in your code, all on one line. That's because, once the constant is declared, its value cannot be altered elsewhere in the code. For example:

```
const char apple = 'a';
```

Constant `apple` is used just like a *char* variable, though its value — the lowercase letter *a* — never changes:

```
printf("Start with letter %c.",apple);
```

The preceding statement outputs this text:

```
Start with letter a.
```

If the code attempts to change a constant, the compiler generates an error message and the program isn't created. This error is the protection afforded by the compiler to the *const* keyword: Constants cannot be changed.

REMEMBER

- Constants must be assigned values on the same line where they're declared.
- You can define a constant of any data type.
- Unlike with a variable, you cannot reassign a constant's value. Otherwise, it's used like any other variable in your code.

TECHNICAL STUFF

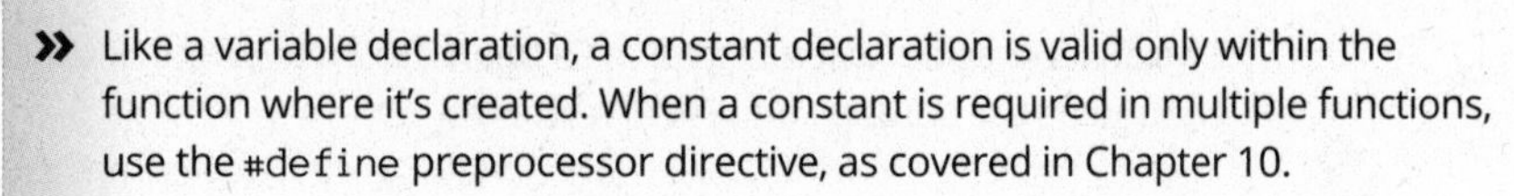

- Like a variable declaration, a constant declaration is valid only within the function where it's created. When a constant is required in multiple functions, use the `#define` preprocessor directive, as covered in Chapter 10.

Putting constants to use

Anytime your code uses a single value over and over (something significant, like the number of rows in a table or the maximum number of items you can stick in a shopping cart), declare the value as a constant.

Exercise 6-15: Rewrite the source code for Listing 6-8. Use a constant `v` to represent the value 3 used in the various *printf()* statements.

Remember that constants are declared just like variables, though the statement begins with the *const* keyword. Review the preceding section, if necessary.

Because I'm not that evil, Listing 6-9 shows my solution for Exercise 6-15.

LISTING 6-9: Preparing for Constant Updates

```
#include <stdio.h>

int main()
{
    const int v = 3;

    printf("The value is %d\n",v);
    printf("And %d is the value\n",v);
    printf("It's not %d\n",v+1);
    printf("And it's not %d\n",v-1);
    printf("No, the value is %d\n",v);
    return(0);
}
```

Integer constant v is declared at Line 5, set to the value 5. Constant v is used in the *printf()* statements exactly as if it were a variable, but replacing the immediate value 5. The advantage to this approach becomes apparent when you tackle Exercise 6-16.

Exercise 6-16: Update your solution for Exercise 6-15 (refer to Listing 6-9) to change the value of variable v to 5. (This is an update to Exercise 6-14.)

The difference between Exercise 6-14 and Exercise 6-16 is that you make only one change in the source code file for the latter.

Exercise 6-17: Modify the source code from Listing 6-7 so that variables a and b are constants.

TIP

You can start out creating a constant. But in my experience, constants tend to evolve: As you work out problems and refine the code, the need for constants becomes apparent. Take advantage of them! If you find yourself using a value consistently in the code, or even a variable when its value doesn't change, declare a constant.

» Reading and writing characters

» Understanding *getchar()* and *putchar()*

» Exploring the *char* variable type

» Reading input with *scanf()*

» Grabbing strings with *fgets()*

Chapter 7

Input and Output

One of the basic functions, if not *the* basic function, of any computing device is input and output. The old I/O (say "eye oh") is also the goal of just about every program. Input is received and processed, and then output is generated. The processing is what makes the program useful. Otherwise, you'd have only input and output, which is essentially the same thing as plumbing.

Character I/O

The simplest type of input and output takes place at the character level: One character goes in; one character comes out. Of course, getting to that point involves a wee bit of programming.

Understanding input and output devices

The C language was born with the Unix operating system. As such, it follows many of the rules for that operating system with regard to input and output. Those rules are pretty solid:

» Input comes from the standard input device, `stdin`.

» Output is sent to the standard output device, `stdout`.

On a computer, the standard input device, `stdin`, is the keyboard. Input can also be redirected by the operating system, so it can come from another device, like a modem or a file.

The standard output device, `stdout`, is the display. Output can be redirected so that it goes to another device, such as a printer or into a file.

REMEMBER

C language functions that deal with input and output access the `stdin` and `stdout` devices. They do not directly read from the keyboard or output to the screen. Well, unless you code your program to do so. (Such coding isn't covered in this book.)

Bottom line: Although your programs can get input from the keyboard and send output to the display, you need to think about C language I/O in terms of `stdin` and `stdout` devices instead. If you forget that, you can get into trouble, which I happily demonstrate later in this chapter.

Fetching characters with *getchar()*

It's time for your code to become more interactive. Consider the source code from Listing 7-1, which uses the *getchar()* function. This function gets (reads) a character from standard input.

LISTING 7-1: It Eats Characters

```
#include <stdio.h>

int main()
{
    int c;

    printf("I'm waiting for a character: ");
    c = getchar();
    printf("I waited for the '%c' character.\n",c);
    return(0);
}
```

The code in Listing 7-1 uses the *getchar()* function at Line 8 to fetch a character from standard input. The character returned from *getchar()* is stored in the `c` integer variable.

Line 9 displays the character stored in `c`. The *printf()* function uses the `%c` placeholder to display single characters.

Exercise 7-1: Create a new program using the source code from Listing 7-1. Build and run.

The *getchar()* function is prototyped in the `stdio.h` header file, which must be included in the source code. In Listing 7-1, it's included already for the definition of the *printf()* function. The *getchar()* function's format is

```
int getchar(void);
```

The function has no arguments, which is why the keyword *void* is set for the function's argument in its definition. In use, the parentheses are always empty.

The *getchar()* function returns an integer value — not a *char* value, despite its being a character I/O function. The integer returned contains a character, which is just how it works.

Exercise 7-2: Edit Line 9 in the source code from Listing 7-1 so that the `%d` placeholder is used instead of `%c`. Build and run.

The value that's displayed when you run the solution to Exercise 7-1 is the character's ASCII code value. The `%d` displays that value because internally the computer treats all information as values. Only when information is displayed as a character does it look like text.

Appendix A lists ASCII code values.

TECHNICAL STUFF

Depending on how the compiler implements it, *getchar()* may be a macro, not a function. A *macro* is a shortcut based on another function. For example, the real function to get characters from standard input may be *getc()*; specifically, when used like this:

```
c = getc(stdin);
```

In this example, *getc()* reads from the standard input device, `stdin`, which is defined in the `stdio.h` header file. The function returns an integer value, which is stored in variable `c`.

Exercise 7-3: Rewrite the source code for Listing 7-1, replacing the *getchar()* statement with the *getc()* example just shown.

Exercise 7-4: Write a program that prompts for three characters; for example:

```
I'm waiting for three characters:
```

Code three consecutive *getchar()* functions to read the characters. Format the result like this:

```
The three characters are 'a', 'b', and 'c'
```

where these characters — `a`, `b`, and `c` — would be replaced by the program's input.

TIP

The program you create in Exercise 7-4 waits for three characters. The Enter key is a character, so if you type **A**, **Enter**, **B**, **Enter**, the three characters are *A*, the Enter key character, and then *B*. This input is valid, but what you probably want to type is something like **ABC** or **PIE** or **LOL** and then press the Enter key.

REMEMBER

Standard input is stream-oriented. As I mention earlier in this chapter, don't expect your C programs to be interactive. Exercise 7-4 is an example of how stream input works; the Enter key doesn't end stream input; it merely rides along in the stream, like any other character.

Using the *putchar()* function

The output counterpart to *getchar()* is the *putchar()* function. It's also prototyped in the `stdio.h` header file. The function serves the purpose of sending a single character to standard output. Here's the format:

```
int putchar(int c);
```

The function's single argument is an integer value, `c` in the preceding line of code. The argument can be a variable or a literal character, as in

```
putchar('v');
```

The function returns the integer value of the character output, though this value need not be checked for simple character output, as shown in Listing 7-2.

LISTING 7-2: Putting *putchar()* to Work

```
#include <stdio.h>

int main()
{
    int ch;
```

```
    printf("Press Enter: ");
    getchar();
    ch = 'H';
    putchar(ch);
    ch = 'i';
    putchar(ch);
    putchar('!');
    return(0);
}
```

This chunk of code uses the *getchar()* function to pause the program at Line 8. The input that's received isn't stored; it doesn't need to be. The compiler doesn't complain when you don't keep the value returned from the *getchar()* function (or from any function).

In Lines 9 through 12, single-character values are assigned to the `ch` variable, which works even though `ch` is an integer variable. The *putchar()* function then displays the changing value of variable `ch`.

In Line 13, *putchar()* displays a character literal, the exclamation point character. Again, the character must be enclosed in single quotes.

Exercise 7-5: Create a new source file using the code shown in Listing 7-2. Build and run the program.

One weird thing about the output is that the final character isn't followed by a newline. That output can look awkward on a text display, so:

Exercise 7-6: Modify the source code from Exercise 7-5 so that the newline character is output after the ! character.

Working with character variables

The *getchar()* and *putchar()* functions work with integers, but that doesn't mean you must shun character variables. When you work with characters in your code, use the *char* data type to store them, as shown in Listing 7-3.

LISTING 7-3: Character Variable Madness

```
#include <stdio.h>

int main()
{
    char a,b,c,d;

    a = 'W';
    b = a + 24;
    c = b + 8;
    d = '\n';
    printf("%c%c%c%c",a,b,c,d);
    return(0);
}
```

Exercise 7-7: Create a new source code file using the code in Listing 7-3. Build and run the program.

The code declares four *char* variables at Line 5. These variables are assigned values in Lines 7 through 10. Line 7 is pretty straightforward. Line 8 uses math to set the value of variable `b` to a specific character, as does Line 9 for variable `c`. (Use Appendix A to look up a character's ASCII code value.) Line 10 uses an escape sequence to set a character's value, something you can't type at the keyboard.

All those `%c` placeholders are stuffed into the *printf()* statement, but the output is, well, surprising.

Exercise 7-8: Modify the code for Listing 7-3 so that variables `b` and `c` are assigned their character values directly using character literals held in single quotes.

Exercise 7-9: Modify the source code again so that *putchar()*, not *printf()*, is used to generate output.

Text I/O, but Mostly I

When character I/O is taken up a notch, it becomes text I/O. The primary text output functions are *puts()* and *printf()*. On the I side of I/O are text input functions, primarily *scanf()* and *fgets()*.

Storing strings

When a program needs text input, it's necessary to create a place to store that text. Right away, you'll probably say, "Golly! That would be a string variable." If you answered that way, I admire your thinking. You're relying upon your knowledge that *text* in C programming is referred to as a *string.*

Alas, you're wrong.

C lacks a string variable type. It does, however, have character variables. Queue up enough of them and you have a string. Or, to put it in programming lingo, you have an *array* of character variables.

Arrays are a big topic, covered in Chapter 12. For now, be open-minded about arrays and strings and soak in the goodness of Listing 7-4.

LISTING 7-4: Stuffing a String into a *char* Array

```
#include <stdio.h>

int main()
{
    char prompt[] = "Press Enter to explode:";

    printf("%s",prompt);
    getchar();
    return(0);
}
```

Line 5 creates an array of *char* variables. An *array* is a gizmo that stores a bunch of the same data types all in a row. The *char* array is named `prompt`, which is immediately followed by empty square brackets. This is the Big Clue that the construction is an array. The array is initialized, via the equal sign, to the text enclosed in double quotes.

The *printf()* statement in Line 7 displays the string stored in the `prompt` array. The `%s` conversion character represents a string.

In Line 8, *getchar()* pauses the program, anticipating the Enter key press. The program doesn't follow through by exploding anything, a task I leave up to you to code at a future date.

Exercise 7-10: Create a new program using the source code from Listing 7-4. Build and run.

Exercise 7-11: Modify the source code from Listing 7-4 so that a single string variable holds two lines of text; for example:

```
Program to Destroy the World
Press Enter to explode:
```

Hint: Refer to Table 4-1, in Chapter 4.

» A string "variable" in C is really a character array.

» You can initialize a string to a *char* array when it's created, similarly to the way you initialize any variable when it's created. The format looks like this:

```
char string[] = "text";
```

In the preceding line, `string` is the name of the *char* array, and `text` is the string assigned to that array.

» You cannot reassign or change a character array's contents by using a direct statement later in the code, such as

```
prompt = "This is just wrong.";
```

Changing a string is possible in C, but you need to know more about arrays, string functions, and especially pointers before you make the attempt. Later chapters in this book cover these topics.

» See Chapter 6 for an introduction to the basic C language data types. The full list of C language data types is found in Appendix D.

Introducing the *scanf()* function

For the input of specific data types, the *scanf()* function comes in handy. It's not a general-purpose input function, and it has some limitations, but it's great for testing code or grabbing values.

In a way, you could argue that *scanf()* is the input version of the *printf()* function. For example, it uses the same conversion characters (the `%` placeholder-things). Because of that, *scanf()* is quite particular about how text is input.

The *scanf()* function is prototyped in the `stdio.h` header file. Here's the format:

```
int scanf(const char *restrict format,...);
```

Scary, huh? Just ignore it for now. Here's my less frightening version of the format:

```
scanf("placeholder",variable);
```

In this version, *`placeholder`* is a conversion character, and *`variable`* is a variable that matches the conversion character's data type. Unless *`variable`* is a string (*char* array), it's prefixed by the `&` operator.

Here are some *scanf()* examples:

```
scanf("%d",&highscore);
```

The preceding statement reads an integer value from standard input into the variable `highscore`, an *int* variable.

```
scanf("%f",&temperature);
```

The preceding *scanf()* statement waits for a floating-point value to be input, which is stored in the `temperature` variable.

```
scanf("%c",&key);
```

In the preceding statement, *scanf()* accepts the first character read from standard input and stores it in the `key` variable.

```
scanf("%s",firstname);
```

REMEMBER

The `%s` placeholder is used to read in text, but only until the first whitespace character is encountered. So a space, a tab, or the Enter key terminates the string. Also, `firstname` is a *char* array, so it need not be prefixed by the `&` operator.

Reading a string with *scanf()*

It's rare to use the *scanf()* function to read a string, because input is terminated at the first whitespace character. Still, it's worth a try.

To use the *scanf()* function to read in a chunk of text, the `%s` conversion character is used — just like in *printf()* but with input instead of output, as shown in Listing 7-5.

LISTING 7-5: *scanf()* Swallows a String

```
#include <stdio.h>

int main()
{
    char firstname[15];

    printf("Type your first name: ");
    scanf("%s",firstname);
    printf("Pleased to meet you, %s.\n",firstname);
    return(0);
}
```

Exercise 7-12: Type the source code from Listing 7-5 into your editor. Save, build, and run.

Line 5 declares a *char* array — a buffer to store a string — named `firstname`. The number in the brackets indicates the size of the array, or the total number of characters that can be stored there. The array isn't assigned a value, so it's created empty. Basically, the statement at Line 5 sets aside storage for up to 15 characters.

The *scanf()* function in Line 8 reads a string from standard input and stores it in the `firstname` array. The `%s` conversion character directs *scanf()* to look for a string as input, just as `%s` is a placeholder for strings in *printf()*'s output.

Exercise 7-13: Modify the source code from Listing 7-5 so that a second string is declared for the person's last name. Prompt the user for their last name as well, and then display both names by using a single *printf()* function.

- The number in the brackets (refer to Line 5 in Listing 7-5) gives the size of the *char* array, or the length of the string, minus *one*. A 15-character array holds a string up to 14 characters long.
- The reason for a *char* array of size *n* holding *n*–1 characters is that all strings in C end with the null character, which is written as escape sequence `\0`. In C, as well as in other programming languages, the null character terminates a string. For literal strings, the compiler automatically appends the `\0` character. For strings obtained from standard input or those you create, you must remember to save one character of storage for the null character.
- The null character is not the same as the NULL pointer constant. Part 4 of this book covers pointers.

WARNING

» The *scanf()* function as used in this chapter is considered insecure by modern coding standards. These exercises are good for learning C and practicing your coding. In the real world, however, programmers avoid using the *scanf()* function for input, because it can easily be exploited.

Reading values with *scanf()*

The *scanf()* function can do more than read strings. It can read in any value specified by a conversion character, as demonstrated in Listing 7-6.

LISTING 7-6: *scanf()* Eats an Integer

```
#include <stdio.h>

int main()
{
    int fav;

    printf("What is your favorite number: ");
    scanf("%d",&fav);
    printf("%d is my favorite number, too!\n",fav);
    return(0);
}
```

In Listing 7-6, the *scanf()* function reads an integer value from standard input. The `%d` conversion character is used, just like *printf()* — indeed, it's used again in Line 9. This character directs *scanf()* to look for an *int* value for variable `fav`.

Exercise 7-14: Write new source code using Listing 7-6. Build and run. Test the program by typing various integer values, positive and negative.

The ampersand (`&`) operator is used to prefix the second argument in the *scanf()* function. The `&` is the address-of operator. It's one of the advanced features in C, related to pointers. I avoid the topic of pointers until Chapter 18, but for now, know that an ampersand must prefix any variable specified in the *scanf()* function. The exception is an array, such as the `firstname` *char* array in Listing 7-5.

Try running the program again, but specify a decimal value, such as 41.9, or type text instead of a number.

The reason you see incorrect output is that *scanf()* is *very* specific. It fetches only the variable type specified by the conversion character. So if you want a floating-point value, you must specify a *float* variable and use the appropriate conversion character; `%f`, in this case.

Exercise 7-15: Modify the source code from Listing 7-6 so that a floating-point number is requested, input, and displayed.

REMEMBER

- You don't need to prefix a *char* array variable with an ampersand in the *scanf()* function; when using *scanf()* to read in a string, just specify the string variable name.
- The *scanf()* function stops reading text input at the first whitespace character, space, tab, or Enter key.

Using *fgets()* for text input

For a general-purpose text input function, one that reads beyond the first whitespace character, I recommend the *fgets()* function. Like most of the other text I/O functions, it's prototyped in the `stdio.h` header file. Here's the format:

```
char * fgets(char *restrict s, int n, FILE *restrict stream);
```

Frightening, no? The *fgets()* function is a file function, which reads text from a file, as in "file get string." That's how programmers talk after an all-nighter.

File functions are covered in Chapter 22, but because the operating system considers standard input like a file, you can use *fgets()* to read text from the keyboard.

Here's a simplified version of the *fgets()* function as it applies to reading text input:

```
fgets(string,size,stdin);
```

In this example, *`string`* is the name of a *char* array, a string variable; *`size`* is the number of characters to input plus *one* for the null character. Effectively, the *`size`* value is the same size as the *char* array. The final argument is the `stdin` constant, the name of the standard input device as defined in the `stdio.h` header file.

Listing 7-7 shows code that uses the *fgets()* function to read input.

LISTING 7-7: The *fgets()* Function Reads a String

```
#include <stdio.h>

int main()
{
    char name[10];

    printf("Who are you? ");
    fgets(name,10,stdin);
    printf("Glad to meet you, %s.\n",name);
    return(0);
}
```

Exercise 7-16: Type the source code from Listing 7-7 into your editor. Build and run.

The *fgets()* function in Line 8 reads text from standard input. The text is stored in the name array, which is set to a maximum of ten characters in Line 5. The value 10 is used as the second argument in the *fgets()* function to ensure that only nine characters are read, one less than the number specified. The final character in the name buffer is the null character, `\0`, which terminates the string.

The last argument in the *fgets()* function is `stdin`, the standard input device. This is the "file" from which input is read.

REMEMBER

The *char* array must have one extra character reserved for the `\0` at the end of a string. Its size must equal the size of input you need — plus one.

Here's how the program runs on my screen:

```
Who are you? Danny Gookin
Glad to meet you, Danny Goo.
```

Only the first nine characters of the text I typed in the first line are displayed. Why only nine? Because the tenth character must be the null character, `\0`, terminating the string. If the *fgets()* function were to read in ten characters instead of nine, the array would overflow and the program could malfunction.

If you type fewer than ten characters, you see that the Enter character is also stored in the string. This effect causes the period at the end of the *printf()* statement to appear alone on the following line:

```
Who are you? Danny
Glad to meet you, Danny
.
```

Exercise 7-17: Change the array size in the source code from Listing 7-7 to a constant value. Set the constant to allow only three characters input.

Exercise 7-18: Redo your solution for Exercise 7-13 so that *fgets()* rather than *scanf()* is used to read in the two strings.

You can read more about the reason *fgets()* is a preferred text input function in the nearby sidebar, "Avoid *fgets()*'s evil sibling *gets()*."

REMEMBER

- The *fgets()* function as presented in this section reads text from the standard input device, not from the keyboard directly.
- The value returned by *fgets()* is the string that was input. In this book's sample code, this return value isn't used, although upon success, the string returned is identical to the information stored in the *fgets()* function's first argument, the *char* array variable.
- Chapter 13 offers more information about strings in C.

WARNING

AVOID *FGETS()*'S EVIL SIBLING *GETS()*

The original C language string-input function was *gets()*, for get-string. It took only a single argument, the character buffer or array into which text input was stored. The problem? No limit was set on this input, which means the buffer could easily overflow and wreak havoc in the computer. In fact, many early computer viruses were written specifically to exploit this weakness in the *gets()* function.

The *gets()* function is still available in the C library, though the compiler scolds you heavily for using it. Further, many compilers add warning text to any program created with the *gets()* function included in the code.

Bottom line: Don't use *gets()* — use *fgets()* instead.

» Using comparison operators

» Adding *else* to the decision

» Creating an *if-else-if-else* structure

» Making logical decisions

» Working with a *switch-case* structure

» Appreciating the ternary operator

Chapter 8
Decision Making

Decision making is the part of programming that makes you think a computer is smart. It's not, of course, but you can fool anyone by crafting your code to carry out directions based on certain conditions or comparisons. The process is easy to understand, but deriving that understanding from the weirdo way it looks in a C program is why this chapter is necessary.

What if?

All human drama is based on disobedience. No matter what the rules, no matter how strict the guidelines, some joker breaks free and the rest is an interesting story. This adventure begins with the simple human concept of "what if." It's the same concept used for decision making in your programs, though in that instance only the word *if* is required.

Making a simple comparison

You make comparisons all the time. What will you wear in the morning? Should you avoid Bill's office because Marjorie says he's "testy" today? And how much longer will you put off going to the dentist? The computer is no different, albeit the comparisons it makes use values, not abstracts, as illustrated in Listing 8-1.

LISTING 8-1: A Simple Comparison

```
#include <stdio.h>

int main()
{
    int a,b;

    a = 6;
    b = a - 2;

    if( a > b)
    {
        printf("%d is greater than %d\n",a,b);
    }
    return(0);
}
```

Exercise 8-1: Create a new program using the source code shown in Listing 8-1. Build and run. Here's the output you should see:

```
6 is greater than 4
```

Fast and obedient, that's what a computer is. Here's how the code works:

Line 5 declares two integer variables: `a` and `b`. The variables are assigned values in Lines 7 and 8; the value of variable `b` is assigned to the value of variable `a` minus 2.

Line 10 makes a comparison:

```
if( a > b)
```

Programmers read this line as, "If *a* is greater than *b*." Or when they're teaching the C language, they say, "If variable `a` is greater than variable `b`." And, no, they don't announce the parentheses.

Lines 11 through 13 form a block that belongs to the *if* statement. The meat in the sandwich is Line 12; the curly brackets don't play a decision-making role, other than hugging the statement at Line 12. When the *if* expression at Line 10 is true, the statement in Line 12 is executed. Otherwise, all statements in the block are skipped.

Exercise 8-2: Edit the source code from Listing 8-1 so that addition instead of subtraction is performed in Line 8. Can you explain the program's output?

Introducing the *if* keyword

The *if* keyword is used to make decisions in your code based upon simple comparisons. Here's the basic format:

```
if(expression)
{
    statement;
}
```

The `expression` is a comparison, a mathematical operation, the result of a function, or some other condition. When the expression is true, the `statements` (or `statement`) enclosed in braces are executed; otherwise, they're skipped.

- The *if* statement's expression need not be mathematical. It can be a function that returns a true or false value, for example:

  ```
  if( ready() )
  ```

  This statement evaluates the return of the *ready()* function. If the function returns a true value, the statements belonging to *if* are executed.
- Any non-zero value is considered true in C. Zero is considered false. So this statement is true:

  ```
  if(1)
  ```

  And this statement is always false:

  ```
  if(0)
  ```
- You know whether a function returns a true or false value by reading the function's documentation, or you can set a true or false return value when writing your own functions.
- You cannot compare strings by using an *if* comparison. Instead, you use specific string comparison functions, which are covered in Chapter 13.
- When only one statement belongs to an *if* comparison, the braces are optional.

TIP

Exercise 8-3: Rewrite the code from Listing 8-1, removing the braces before and after Line 12. Build and run to ensure that it still works.

Comparing values in various ways

The C language employs a small platoon of mathematical comparison operators. I've gathered the bunch in Table 8-1 for your perusal.

TABLE 8-1 C Language Comparison Operators

Operator	Pronunciation	Example	True When
!=	Not equal to	a != b	a is not equal to b
<	Less than	a < b	a is less than b
<=	Less than or equal to	a <= b	a is less than or equal to b
==	Is equal to	a == b	a is equal to b
>	Greater than	a > b	a is greater than b
>=	Greater than or equal to	a >= b	a is greater than or equal to b

Comparisons in C read from left to right, so you read `a >= b` as "a is greater than or equal to `b`." Also, the order is important: Both `>=` and `<=` must be written in that order, as must the `!=` (not equal) operator. The `==` operator can be written either way.

Listing 8-2 shows *if* expressions using the less-than and greater-than operators.

LISTING 8-2: Values Are Compared

```
#include <stdio.h>

int main()
{
    int first,second;

    printf("Input the first value: ");
    scanf("%d",&first);
    printf("Input the second value: ");
    scanf("%d",&second);
```

```
    puts("Evaluating...");
    if(first<second)
    {
        printf("%d is less than %d\n",first,second);
    }
    if(first>second)
    {
        printf("%d is greater than %d\n",first,second);
    }
    return(0);
}
```

Exercise 8-4: Create a new project by using the source code shown in Listing 8-2. Build and run.

The most common expression *if* examines is probably the double equal sign. It may look odd to you. The == operator isn't the same as the = operator. The = operator is the *assignment operator,* which sets values. The == operator is the *comparison operator,* which checks to see whether two values are equal.

TIP

I pronounce == as "is equal to."

Exercise 8-5: Add a new section to the source code from Listing 8-2 that makes a final evaluation on whether both variables are equal to each other.

Exercise 8-6: Type the source code from Listing 8-3 into a new source code file. Build and run.

LISTING 8-3: Get "Is Equal To" into Your Head

```
#include <stdio.h>

int main()
{
    const int secret = 17;
    int guess;

    printf("Can you guess the secret number: ");
    scanf("%d",&guess);
    if(guess==secret)
    {
        puts("You guessed it!");
        return(0);
```

(continued)

LISTING 8-3: *(continued)*

```
    }
    if(guess!=secret)
    {
        puts("Wrong!");
        return(1);
    }
}
```

Take note of the value returned by the program — either 0 for a correct answer or 1 for a wrong answer. If you're using Code::Blocks or another IDE, you can see this value in the output window.

Knowing the difference between = and ==

One of the most common mistakes made by every C language programmer — beginner and pro alike — is using a single equal sign instead of a double in an *if* expression. To wit, I offer Listing 8-4.

LISTING 8-4: **Always True**

```
#include <stdio.h>

int main()
{
    int a;

    a = 5;

    if(a=-3)
    {
        printf("%d equals %d\n",a,-3);
    }
    return(0);
}
```

Exercise 8-7: Use the source code shown from Listing 8-4 to create a new program. Ignore any warnings and run the program.

The output may puzzle you. What I see is this:

```
-3 equals -3
```

That's true, isn't it? But what happened?

In Line 9, variable a is assigned the value -3. Because this expression is inside the parentheses, it's evaluated first. The result of a variable assignment in C is always true for any non-zero value.

Exercise 8-8: Edit the source code from Listing 8-4 so that a double equal sign, or "is equal to," is used instead of the single equal sign in the *if* comparison.

Forgetting where to put the semicolon

Listing 8-5 is based upon Listing 8-4, taking advantage of the fact that C doesn't require a single statement belonging to an *if* comparison to be lodged as a block hugged by curly brackets.

LISTING 8-5: Semicolon Boo-Boo

```
#include <stdio.h>

int main()
{
    int a,b;

    a = 5;
    b = -3;

    if(a==b);
        printf("%d equals %d\n",a,b);
    return(0);
}
```

Exercise 8-9: Carefully type the source code from Listing 8-5. Pay special attention to Line 10. Ensure that you type it in exactly, with the semicolon at the end of the line. Ignore any warnings. Build and run the program.

Here's the output I see:

```
5 equals -3
```

The problem here is a common one, a mistake made by just about every C programmer: The trailing semicolon in Listing 8-5 (Line 10) tells the program that the *if* statement has nothing to do when the condition is true. That's because a single semicolon is a complete statement in C, albeit a null statement. To wit:

```
if(condition)
    ;
```

This construction is basically the same as Line 10 in Listing 8-5. Be careful not to make the same mistake — especially when you type code a lot and you're used to ending a line with a semicolon.

Multiple Decisions

Not every decision is a clean-cut, yes-or-no proposition. Exceptions happen all the time. C provides a few ways to deal with those exceptions, allowing you to craft code that executes based on multiple possibilities.

Making more-complex decisions

For the either-or type of comparisons, the *if* keyword has a companion — *else*. Together, they work like this:

```
if(expression)
{
    statement(s);
}
else
{
    statement(s);
}
```

When the `expression` is true in an *if-else* structure, the statements belonging to *if* are executed; otherwise, the statements belonging to *else* are executed. It's an either-or type of decision.

Listing 8-6 is an update of sorts to the code shown in Listing 8-1. The single *if* structure has been replaced by *if-else*. When the *if* comparison is false, the statement belonging to *else* is executed.

LISTING 8-6: **An *if-else* Comparison**

```
#include <stdio.h>

int main()
{
    int a,b;

    a = 6;
    b = a - 2;

    if( a > b)
    {
        printf("%d is greater than %d\n",a,b);
    }
    else
    {
        printf("%d is not greater than %d\n",a,b);
    }
    return(0);
}
```

Exercise 8-10: Type the source code for Listing 8-6 into a new project. Compile and run.

Exercise 8-11: Modify the source code so that the user gets to input the value of variable `b`.

Exercise 8-12: Modify the source code from Listing 8-3 so that an *if-else* structure replaces that ugly *if-if* thing. (***Hint:*** The best solution changes only one line of code.)

Adding a third option

Not every decision made in a program is either-or. Sometimes, you find yourself in need of an either-or-or-and-then type of thing. No word exists in English to describe such a structure, but it exists in C. It looks like this:

```
if(expression)
{
    statement(s);
}
else if(expression)
```

```
{
    statement(s);
}
else
{
    statement(s);
}
```

When the first *expression* proves false, the *else if* statement makes another test. If that *expression* proves true, its statements are executed. When neither condition is true, the statements belonging to the final *else* are executed.

Exercise 8-13: Using the source code from Listing 8-2 as a base, create an *if-if else-else* structure that handles three conditions. The first two conditions are specified in Listing 8-2, and you need to add the final possibility using a structure similar to the one shown in this section.

C has no limit on how many *else if* statements you can add to an *if* decision process. Your code could show an *if*, followed by three *else-if* conditions, and a final *else*. This process works, though it may not be the best approach. See the later section "Making a multiple-choice selection," for a better way.

Multiple Comparisons with Logic

Some comparisons are more complex than those presented by the simple operators illustrated earlier, in Table 8-1. For example, consider the following math-thingie:

```
-5 <= x <= 5
```

In English, this statement means that x represents a value between −5 and 5, inclusive. That's not a C language *if* comparison, but it can be, when you employ logical operators.

Building a logical comparison

It's possible to load two or more comparisons into a single *if* statement. The results are then compared by using a logical operator. When the entire thing is true, the *if* condition is considered true, as shown in Listing 8-7.

LISTING 8-7: Logic Is a Tweeting Bird

```
#include <stdio.h>

int main()
{
    int coordinate;

    printf("Input target coordinate: ");
    scanf("%d",&coordinate);
    if( coordinate >= -5 && coordinate <= 5 )
    {
        puts("Close enough!");
    }
    else
    {
        puts("Target is out of range!");
    }
    return(0);
}
```

Two comparisons are made by the *if* statement condition in Line 9. That statement reads like this: "If the value of variable *coordinate* is greater than or equal to −5 *and* less than or equal to 5."

Exercise 8-14: Create a new project using the source code from Listing 8-7. Build the program. Run the code a few times to test how well it works.

Adding some logical operators

The C language logical comparison operators are shown in Table 8-2. These operators can be used in an *if* expression when two or more conditions must be met.

TABLE 8-2 Logical Comparison Operators

Operator	Name	True When
&&	and	Both comparisons are true
||	or	Either comparison is true
!	not	The item is false

Listing 8-7 uses the `&&` operator as a logical AND comparison. Both conditions specified must be true for the *if* statement to consider everything in the parentheses to be true.

Exercise 8-15: Modify the source code from Listing 8-7 so that a logical OR operation is used to make the condition true when the value of variable `coordinate` is less than −5 or greater than 5.

Exercise 8-16: Create source code for a program that asks the popular question, "Do you want to continue (Y/N)?" Process single-character input, testing for Y or N, either upper- or lowercase. Ensure that the program responds properly when neither a Y nor N is input.

- » Logical operations are often referred to by using all caps: AND, OR. That separates them from the normal words *and* and *or.*
- » The logical AND is represented by two ampersands: &&. Say "and."
- » The logical OR is represented by two pipe, or vertical-bar, characters: `||`. Say "or."
- » The logical NOT is represented by a single exclamation point: !. Say "not!"
- » The logical NOT prefixes a value, or an expression in parentheses, to reverse the results, transforming False into True and True into False.

The Old *Switch Case* Trick

Piling up a tower of *if* and *if-else* statements can be effective, but it's not the best way to walk through some multiple-choice decisions. An alternative offered in the C language is known as the *switch-case* structure.

Making a multiple-choice selection

The *switch-case* structure allows you to code decisions in a C program based on a single value. It's the multiple-choice selection statement, as shown in Listing 8-8.

LISTING 8-8: Multiple Choice

```
#include <stdio.h>

int main()
{
    int code;

    printf("Enter the error code (1-3): ");
    scanf("%d",&code);
    switch(code)
    {
        case 1:
            puts("Drive Fault, not your fault.");
            break;
        case 2:
            puts("Illegal format, call a lawyer.");
            break;
        case 3:
            puts("Bad filename, spank it.");
            break;
        default:
            puts("That's not 1, 2, or 3");
    }
    return(0);
}
```

Exercise 8-17: Create a new program using the code from Listing 8-8. Just type it in; I describe it later. Build it. Run it a few times, trying various values to see how it responds.

Examine the source code in your editor, where you can reference the line numbers mentioned in the following paragraphs.

The *switch-case* structure starts at Line 9 with the *switch* statement. The item it evaluates is enclosed in parentheses. Unlike an *if* statement, *switch* eats only a single value, not a comparison. In Line 9, the value is an integer that the user types (read in Line 8).

The *case* part of the structure is enclosed in curly brackets, between Lines 10 and 22. A *case* statement shows a single value, such as 1 in Line 11. The value is followed by a colon.

The value specified by each *case* statement is compared with the item specified in the *switch* statement. If the values are equal, the statements belonging to *case* are executed. If not, they're skipped and the next *case* value is compared.

The *break* keyword stops program flow through the *switch-case* structure. Program flow resumes after the *switch-case* structure's final curly bracket, which is Line 23 in Listing 8-8.

After the final comparison, the *switch-case* structure uses a *default* item, shown in Line 20. This item's statements are executed when none of the *case* comparisons matches. The *default* item need not be specified, though in Listing 8-8 it handles out-of-range values.

Exercise 8-18: Construct a program using source code similar to Listing 8-8, but make the input the letters *A*, *B*, and *C*. You might want to review Chapter 7 to see how single characters are specified in the C language.

REMEMBER

- The comparison being made in a *switch-case* structure is between the value specified in *switch*'s parentheses and the values that follow each *case* keyword. When the comparison is true, meaning that both values are equal to each other, the statements belonging to *case* are executed.
- The *break* keyword disrupts program flow. It can be used in an *if* structure as well, but mostly it's found in loops. See Chapter 9.
- Specify a *break* after a *case* comparison's statements so that the rest of the structure isn't executed. See the later section "Taking no breaks."

Understanding the *switch-case* structure

And now — presenting the most complex thing in C. Seriously, you'll find more rules and structure with *switch-case* than just about any other construct in C. Here's the skeleton:

```
switch(expression)
{
    case value1:
        statement(s);
        break;
    case value2:
        statement(s);
        break;
    case value3:
        statement(s);
```

```
        break;
    default:
        statement(s);
}
```

The *switch* item introduces the structure, which follows and is enclosed by a set of curly brackets. The structure must contain at least one *case* statement, though more than one *case* statement is required to make the thing useful.

The *switch* statement contains an `expression` in parentheses. That expression must evaluate to a single value. It can be a variable, a value returned from a function, or a mathematical operation.

A *case* statement is followed by an immediate value and then a colon. Following this statement are one or more statements. These statements are executed when the immediate value following *case* matches the *switch* statement's expression. Otherwise, the statements are skipped and the next *case* statement is evaluated.

The *break* keyword is used to flee the *switch-case* structure. Otherwise, program execution cascades through the structure.

The *default* item ends the *switch-case* structure. It contains statements that are executed when none of the *case* statements match. If nothing is left to do, you may omit the *default* item from the *switch-case* structure.

REMEMBER

The *case* portion of a *switch-case* structure doesn't make an evaluation. If the code needs multiple comparisons, use a multiple *if-else* type of structure instead.

Taking no breaks

It's possible to construct a *switch-case* structure with no *break* statements. Such a thing can even be useful under special circumstances, as shown in Listing 8-9.

LISTING 8-9: Meal Plan Decisions

```
#include <stdio.h>

int main()
{
    char choice;

    puts("Meal Plans:");
    puts("A - Breakfast, Lunch, and Dinner");
```

(continued)

LISTING 8-9: *(continued)*

```
    puts("B - Lunch and Dinner only");
    puts("C - Dinner only");
    printf("Your choice: ");
    scanf("%c",&choice);

    printf("You've opted for ");
    switch(choice)
    {
        case 'A':
            printf("Breakfast, ");
        case 'B':
            printf("Lunch and ");
        case 'C':
            printf("Dinner ");
        default:
            printf("as your meal plan.\n");
    }
    return(0);
}
```

Exercise 8-19: Create a new program using the source code from Listing 8-9. Build and run.

Exercise 8-20: If you understand how *case* statements can fall through, modify Exercise 8-19 so that both upper- and lowercase letters are evaluated in the *switch-case* structure.

The Weird ?: Decision Thing

TECHNICAL STUFF

I have one last oddball decision-making tool to throw at you in this long, decisive chapter. It's perhaps the most cryptic of the decision-making tools in C, a favorite of programmers who enjoy obfuscating their code. Witness Listing 8-10.

LISTING 8-10: And Then It Gets Weird

```
#include <stdio.h>

int main()
{
    int a,b,larger;

    printf("Enter value A: ");
    scanf("%d",&a);
    printf("Enter different value B: ");
    scanf("%d",&b);

    larger = (a > b) ? a : b;
    printf("Value %d is larger.\n",larger);
    return(0);
}
```

Specifically, you want to look at Line 12, which I'm showing here as though it isn't ugly enough inside Listing 8-10:

```
larger = (a > b) ? a : b;
```

Exercise 8-21: Create a project using the source code from Listing 8-10. Build and run just to prove that the weirdo `?:` thing works.

Officially, `?:` is known as a *ternary* operator: It's composed of three parts: a comparison, and then value-if-true and value-if-false. Written in plain, hacker English, the statement looks like this:

```
result = expression ? if_true : if_false;
```

The statement begins with an expression. Anything you'd stuff into an *if* statement's parentheses works, as do all operators, mathematical and logical. I typically enclose the expression in parentheses, though this isn't a requirement.

When *`expression`* is true, the *`if_true`* portion of the statement is evaluated and its value is stored in the *`result`* variable. Otherwise, the *`if_false`* solution is stored. Oh, and *`result`* need not be a variable; the value generated by the ternary operator can also be used immediately in a function.

Exercise 8-22: Rewrite the source code form Listing 8-10 using an *if-else* structure to carry out the decision and result from the `?:` ternary operator in Line 12.

» Exploring the *for* loop

» Creating nested *for* loops

» Working a *while* loop

» Using a *do while* loop

» Avoiding the endless loop

Chapter 9
Loops, Loops, Loops

Programs love to do things over and over, mirthfully so. They never complain, they never tire. In fact, they'll repeat things forever unless you properly code instructions for when to stop. Indeed, the loop is a basic programming concept. Do it well. Do it well. Do it well.

A Little Déjà Vu

A *loop* is a section of code that repeats. How often? That depends on how you write the loop. As an overview, a loop involves three things:

- Initialization
- One or more statements that repeat
- An exit

The *initialization* sets up the loop, usually specifying a condition upon which the loop begins or is activated. For example, "Start the counter at 1."

The statements that repeat are contained as a block in curly brackets. They continue to be executed, one after the other, until the exit condition is met.

The *exit condition* determines when the loop stops. Either it's a condition that's met, such as "Stop when the counter equals 10," or the loop can stop when a *break* statement is encountered. The program execution continues with the next statement after the loop's final curly bracket.

WARNING

Having an exit condition is perhaps the most important part of a loop. Without it, the loop repeats forever in a condition called an *endless loop*. See the later section "Looping endlessly."

The C language features two looping keywords: *for* and *while*. Assisting the *while* keyword is the *do* keyword. The *goto* keyword can also be used for looping, though it's heavily shunned.

The Thrill of *for* Loops

A *loop* is a group of statements that repeat. You choose a set number of iterations, or the number of repeats can be based on a value. Either way, the *for* keyword helps set up a basic type of loop.

Doing something x number of times

It's entirely possible, and even a valid solution, to write source code that displays the same line of text ten times. You could copy-and-paste a *printf()* statement to do the job. Simple, but it's not a loop, which is shown in Listing 9-1.

LISTING 9-1: Write That Down Ten Times!

```
#include <stdio.h>

int main()
{
    int x;

    for(x=0; x<10; x=x+1)
    {
        puts("Sore shoulder surgery");
    }
    return(0);
}
```

Exercise 9-1: Create a new program using the source code from Listing 9-1. Type everything carefully, especially Line 7. Build and run.

As output, the program coughs up the tongue-twister *Sore shoulder surgery* ten times, in ten lines of text. The key, of course, is in Line 7, the *for* statement. That statement directs the program to repeat the statement(s) in curly brackets a total of ten times.

Exercise 9-2: Using the source code from Listing 9-1 again, replace the value 10 in Line 7 with the value 20. Build and run.

Introducing the *for* loop

The *for* loop is usually the first type of loop you encounter when you learn to program. It looks complex, but that's because it's doing everything required of a loop in a single statement:

```
for(initialization; exit_condition; repeat_each)
```

Here's how it works:

`initialization` is a C language expression that's evaluated at the start of the loop. Most often, it's where the variable that counts the loop's iterations is initialized.

`exit_condition` is the test upon which the loop stops. In a *for* loop, the statements continue to repeat until the exit condition is true. The expression used for the *exit_condition* is most often a comparison, like something you'd find in an *if* statement.

`repeat_each` is an expression that's executed once every iteration. It's normally an operation affecting the initialization variable at the first part of the *for* statement.

The *for* statement is followed by a block of one or more statements held in curly brackets:

```
for(x=0; x<10; x=x+1)
{
    puts("Sore shoulder surgery");
}
```

You can omit the brackets when only one statement is specified:

```
for(x=0; x<10; x=x+1)
    puts("Sore shoulder surgery");
```

In this *for* statement, and from Listing 9-1, the first expression is initialization:

```
x=0
```

The value of the *int* variable x is set to 0. In C programming, you start counting with 0, not with 1. The advantages of doing so are presented throughout this book.

The second expression sets the loop's exit condition:

```
x<10
```

As long as the value of variable x is less than 10, the loop repeats. Once this expression is false, the loop stops. The result is that the loop repeats ten times. That's because x starts at 0, not at 1.

Finally, the third expression repeats for each iteration of the loop:

```
x=x+1
```

Every time the loop's statements are executed, the value of variable x is increased by 1. The preceding statement reads, "Variable x equals the value of variable x, plus 1." Because C evaluates the right side of the equation first, nothing is goofed up. So if currently the value of x is 5, the new value of x would be 6.

All told, I read the expression this way:

```
for(x=0; x<10; x=x+1)
```

"For *x* starts at 0, while *x* is less than 10, add 1 to *x*."

Listing 9-2 shows another example of a simple *for* loop. It displays values from -5 through 5.

LISTING 9-2: Counting with a Loop

```
#include <stdio.h>

int main()
{
    int count;

    for(count=-5; count<6; count=count+1)
    {
        printf("%d\n",count);
    }
    return(0);
}
```

Exercise 9-3: Enter the source code from Listing 9-2 into the editor. Save, build, and run.

Exercise 9-4: Create a new project using the source code from Listing 9-2 as a starting point. Display the values from 11 through 19. Separate each value by a tab character, \t. Use the <= sign for the comparison that ends the loop. Clean up the display by adding a final newline character when the loop is done.

TECHNICAL STUFF

- The *for* statement uses two semicolons, not commas, to separate each item. Even so:
- It's possible to specify two conditions in a *for* statement by using commas. This setup is rather rare, so don't let it throw you. See the later section "Adding multiple *for* loop conditions" for details.

Counting with the *for* statement

You'll use the *for* statement quite frequently in your coding travels. Listing 9-3 shows another counting variation.

LISTING 9-3: Counting by Two

```
#include <stdio.h>

int main()
{
    int duo;

    for(duo=2;duo<=100;duo=duo+2)
    {
        printf("%d\t",duo);
    }
    putchar('\n');
    return(0);
}
```

Exercise 9-5: Create a new project using Listing 9-3 as your source code. Compile and run.

The program's output displays even values from 2 through 100. The value 100 is displayed because the "while true" condition in the *for* statement uses <= (less than or equal to). The variable duo counts by two because of this expression:

```
duo=duo+2
```

In Line 9, the *printf()* function uses \t to display tabs (though the numbers may not line up perfectly on an 80-column display). Also, the *putchar()* function kicks in a newline character at Line 11.

Exercise 9-6: Modify the source code from Listing 9-3 so that the output starts at the number 3 and displays multiples of 3 all the way up to 100.

Exercise 9-7: Create a program that counts backward from 25 to 0.

Looping letters

Listing 9-4 shows another way to "count" using a *for* loop.

LISTING 9-4: Counting by Letter

```
#include <stdio.h>

int main()
{
    char alphabet;

    for(alphabet='A';alphabet<='Z';alphabet=alphabet+1)
    {
        printf("%c",alphabet);
    }
    putchar('\n');
    return(0);
}
```

Before you type the source code from Listing 9-4, can you guess what the output might be? Does it make sense to you?

Exercise 9-8: Use the source code from Listing 9-4 to create a new program. Build and run.

Exercise 9-9: Modify your solution to Exercise 9-8, changing the *printf()* function in Line 9 so that the `%d` placeholder is used instead of `%c`.

REMEMBER

Computers see characters as numbers. Only when numbers are displayed and they fall in the ASCII code range for characters do characters appear. (See Appendix A for the list of ASCII character codes.)

Exercise 9-10: Using Listing 9-4 as your inspiration, write a *for* loop that "counts" backward from z (lowercase Z) to a (lowercase A).

Nesting *for* loops

One thing you can stick inside a *for* loop is another *for* loop. It may seem crazy to loop within a loop, but it's a common practice. The official jargon is *nested loop.* Listing 9-5 shows an example.

LISTING 9-5: A Nested Loop

```
#include <stdio.h>

int main()
{
    int alpha,code;

    for(alpha='A';alpha<='G';alpha=alpha+1)
    {
        for(code=1;code<=7;code=code+1)
        {
            printf("%c%d\t",alpha,code);
        }
        putchar('\n');          /* end a line of text */
    }
    return(0);
}
```

All the indents are designed to make the code more readable. They show which statements belong to which *for* loop because they line up at the same tab stop.

Line 7 in Listing 9-5 begins the first, outer *for* loop. It counts from letters A to G. It also contains the second, inner *for* loop and a *putchar()* function on Line 13. That function helps organize the output into rows by spitting out a newline after each row is displayed.

The *printf()* function in Line 11 displays the program's output, specifying the outer loop value, `alpha`, and the inner loop value, `code`. The `\t` escape sequence separates the output.

Exercise 9-11: Type the source code from Listing 9-5 into your editor. Build and run.

Here's the output I see on my computer:

```
A1  A2  A3  A4  A5  A6  A7
B1  B2  B3  B4  B5  B6  B7
C1  C2  C3  C4  C5  C6  C7
D1  D2  D3  D4  D5  D6  D7
E1  E2  E3  E4  E5  E6  E7
F1  F2  F3  F4  F5  F6  F7
G1  G2  G3  G4  G5  G6  G7
```

A triple nested loop contains three *for* statements, which continues the cascade shown in Listing 9-5. As long as you can match up the curly brackets with each *for* statement (and that's easy, thanks to modern text editors), it's something you can accomplish quite readily.

Exercise 9-12: Write a 3-letter acronym-generating program. The program's output lists all 3-letter combinations from AAA through ZZZ, spewed out each on a line by itself.

TECHNICAL STUFF

I wrote a program similar to the solution to Exercise 9-12 as one of my first programming projects. The computers in those days were so slow that the output took about ten seconds to run. On today's computers, the output is nearly instantaneous.

The Joy of the *while* Loop

Another popular looping keyword in C is *while*. It has a companion, *do*, so programmers refer to this type of loop as either *while* or *do while*. The C language is missing the *do-whacka-do* type of loop.

Structuring a *while* loop

The C language *while* loop is a lot easier to look at than a *for* loop, but it involves more careful setup and preparation. Basically, it goes like this:

```
while(expression)
{
    statement(s);
}
```

The *expression* is a value or a comparison or any of a number of other things that results in a true/false condition, just like you'd find in an *if* statement. The *expression* is evaluated each time the loop repeats. As long as it's true ("while" it's true), the loop spins and its statements continue to execute.

TIP

Because the expression is evaluated at the start of the loop, the loop must be initialized before the *while* statement, as shown in Listing 9-6.

So how does a *while* loop end? The termination must be triggered within the loop's statements. Usually, one of the statements affects the expression that's evaluated, causing it to be false.

After the *while* loop is done, program execution continues with the next statement after the final curly bracket.

A *while* loop can also forgo the curly brackets when it has only one statement:

```
while(expression)
    statement;
```

LISTING 9-6: The *while* Version of Listing 9-1

```
#include <stdio.h>

int main()
{
    int x;

    x=0;
    while(x<10)
    {
        puts("Sore shoulder surgery");
        x=x+1;
    }
    return(0);
}
```

The *while* loop demonstrated in Listing 9-6 has three parts:

- The initialization takes place on Line 7, where variable x is set equal to 0.
- The loop's exit condition is contained within the *while* statement's parentheses, as shown in Line 8.
- The item that iterates the loop is found on Line 11, where variable x is increased in value.

Exercise 9-13: Create a new program using the source code from Listing 9-6. Build and run.

Exercise 9-14: Change Line 7 in the source code so that variable x is assigned the value 13. Build and run. Can you explain the output?

Exercise 9-15: Write a program that uses a *while* loop to output values from −5 through 5, using an increment of 0.5.

Using the *do while* loop

The *do while* loop can be described as an upside-down *while* loop. This description is true, especially when you look at the thing's structure:

```
do
{
    statement(s);
} while (condition);
```

As with a *while* loop, the initialization must take place before entering the loop, and one of the loop's statements should affect the condition so that the loop exits. The *while* statement, however, appears after the last curly bracket. The *do* statement begins the structure.

Because of its inverse structure, the major difference between a *while* loop and a *do while* loop is that the *do while* loop is always executed at least once. So you can best employ this type of loop when you need to ensure that the statements execute. Likewise, avoid *do while* when you don't want the statements to execute unless the condition is true. An interesting *do while* loop is shown in Listing 9-7.

LISTING 9-7: A Fibonacci Sequence

```
#include <stdio.h>

int main()
{
    int fibo,nacci;

    fibo=0;
    nacci=1;

    do
    {
        printf("%d ",fibo);
        fibo=fibo+nacci;
        printf("%d ",nacci);
        nacci=nacci+fibo;
    } while( nacci < 300 );

    putchar('\n');
    return(0);
}
```

Exercise 9-16: Type the source code from Listing 9-7 into the editor. Mind your typing! The final *while* statement (refer to Line 16) must end with a semicolon; otherwise, the compiler gets all huffy on you.

Here's the output:

```
0 1 1 2 3 5 8 13 21 34 55 89 144 233
```

The loop begins at Lines 7 and 8, where the variables are initialized.

Lines 12 through 15 calculate the Fibonacci values. Two *printf()* functions display the values.

The loop ends on Line 16, where the *while* statement makes its evaluation. As long as variable `nacci` is less than `300`, the loop repeats. You can adjust this value higher to direct the program to output more Fibonacci numbers.

On Line 18, the *putchar()* statement cleans up the output by adding a newline character.

Exercise 9-17: Redo your solution to Exercise 9-14 as a *do while* loop.

Loopy Stuff

I could go on and on about loops all day, repeating myself endlessly! Before moving on, however, I'd like to go over a few looping tips and pratfalls. These things you should know before you get your official *For Dummies* Looping Programmer certificate.

Looping endlessly

Beware the endless loop!

WARNING

When a program enters an endless loop, it either spews output over and over without end or it sits there tight and does nothing. Well, it's doing what you ordered it to do, which is to spin forever. Sometimes, this setup is done on purpose, but mostly it happens because of programmer error. And with the way loops are set up in C, it's easy to unintentionally loop *ad infinitum*.

Listing 9-8 illustrates a common endless loop, which is a programming error, not a syntax error.

LISTING 9-8: A Common Way to Make an Endless Loop

```
#include <stdio.h>

int main()
{
    int x;

    for(x=0;x=10;x=x+1)
    {
        puts("What are you lookin' at?");
    }
    return(0);
}
```

The problem with the code in Listing 9-8 is that the *for* statement's exit condition is always true: `x=10`. Read it again if you didn't catch it the first time, or just do Exercise 9-18.

Exercise 9-18: Type the source code for Listing 9-8. Save and build, ignoring any warnings. Run.

The program runs — infinitely.

TIP

- To break out of an endless loop, press Ctrl+C on the keyboard. This trick works only for console programs, and it may not always work. If it doesn't, you need to kill the process run amok, which is something I don't have time to explain in this book.
- Endless loops are also referred to as *infinite loops*.

Looping endlessly but on purpose

Occasionally, a program needs an endless loop. For example, a microcontroller may load a program that runs as long as the device is on. When you set up such a loop on purpose in C, one of two statements is used:

```
for(;;)
```

I read this statement as "for ever." With no items in the parentheses, but still with the required two semicolons, the *for* loop repeats eternally — even after the cows come home. Here's the *while* loop equivalent:

```
while(1)
```

The value in the parentheses doesn't necessarily need to be 1; any true or non-zero value works. When the loop is endless on purpose, however, most programmers set the value to 1 simply to self-document that they know what's up.

You can see an example of an endless loop on purpose in the next section.

Breaking out of a loop

Any loop can be terminated instantly — including endless loops — by using a *break* statement within the loop's repeating group of statements. When *break* is encountered, looping stops and program execution picks up with the next statement after the loop's final curly bracket. Listing 9-9 demonstrates the process.

LISTING 9-9: Get Me Outta Here!

```
#include <stdio.h>

int main()
{
    int count;

    count = 0;
    while(1)
    {
        printf("%d, ",count);
        count = count+1;
        if( count > 50)
            break;
    }
    putchar('\n');
    return(0);
}
```

The *while* loop at Line 8 is configured to go on forever, but the *if* test at Line 12 can stop it: When the value of count is greater than 50, the *break* statement (refer to Line 13) is executed and the loop halts.

Exercise 9-19: Build and run a new project using the source code from Listing 9-9.

Exercise 9-20: Rewrite the source code from Listing 9-9 so that an endless *for* loop is used instead of an endless *while* loop.

REMEMBER

You don't need to construct an endless loop to use the *break* statement. You can break out of any loop. When you do, execution continues with the first statement after the loop's final curly bracket.

Adding multiple *for* loop conditions

A common mistake in a *for* loop is to use commas instead of semicolons to separate the parts. But commas are allowed in the *for* loop's parentheses, just not to replace the semicolons. Instead, they are used to specify multiple initialization operations as well as multiple looping conditions. Listing 9-10 shows how you can get nutty with commas in a *for* statement.

LISTING 9-10: Crowded in Here

```
#include <stdio.h>

int main()
{
    int a;
    char c;

    for( a=1,c='Z'; a<5; a=a+1,c=c-1 )
        printf("%d%c\n",a,c);
    return(0);
}
```

Line 8 shows a *for* statement with two initializations and two expressions that take place each time the loop repeats. Both pairs are separated by a comma. The semicolons still group each of the three parts in a *for* loop.

First, variable a is initialized to 1 and variable c is initialized to the letter Z. The loop spins as long as the value of variable a is less than 5. And after each statement repeats, the value of variable a is increased by 1 and the value of variable c is decreased by 1.

Exercise 9-21: Type the source code from Listing 9-10 into your editor. Build and run.

Screwing up a loop

I know of two common ways to mess up a loop. These trouble spots crop up for beginners and pros alike. The only way to avoid these spots is to keep a keen eye so that you can spot 'em quick.

The first goof-up is specifying a condition that can never be met; for example:

```
for(x=1;x==10;x=x+1)
```

In the preceding line, the exit condition is false before the loop spins once, so the loop is never executed. This error is almost as insidious as using an assignment operator (a single equal sign) instead of the "is equal to" operator (as just shown).

Another common mistake is misplacing the semicolon, as in

```
for(x=1;x<14;x=x+1);
{
    puts("Sore shoulder surgery");
}
```

Because the first line, the *for* statement, ends in a semicolon, the line is the entire loop. The empty code repeats 13 times, which is what the *for* statement dictates. The *puts()* statement is then executed once.

WARNING

Rogue semicolons can be frustrating!

The problem is worse with *while* loops because the *do while* structure requires a semicolon after the final *while* statement. In fact, forgetting that particular semicolon is also a source of woe. For a traditional *while* loop, you don't do this:

```
while(x<14);
{
    puts("Sore shoulder surgery");
}
```

Most compilers catch these rogue semicolon issues with a warning. Still, be on the lookout, as Listing 9-11 demonstrates.

LISTING 9-11: A *for* Loop with No Body

```
#include <stdio.h>

int main()
{
```

```
    int x;

    for(x=0;x<10;x=x+1,printf("%d\n",x))
        ;
    return(0);
}
```

In the example shown in Listing 9-11, the semicolon is placed on the line after the *for* statement at Line 8. This location shows deliberate intent.

As with the example from the preceding section, two items are placed in the *for* statement's parentheses, both separated by a comma.

Exercise 9-22: Type the source code from Listing 9-11 into your editor. Build and run.

REMEMBER

Though you can load up items in a *for* statement's parentheses, it's rare and not recommended for readability's sake.

WARNING

AVOID *GOTO* HELL

The third looping statement is the most despised and the lowest-of-the-low C language keywords. It's *goto*, which is pronounced "go to," not "gotto." It directs program execution to another line in the source code, a line tagged by label. Here's an example:

```
here:
    puts("This is a type of loop");
goto here;
```

As this chunk of code executes (from the top down), the `here` label is ignored. The *puts()* function comes next, and, finally, *goto* redirects program flow back up to the `here` label. Everything repeats. Everything works. But it's just darn ugly.

Most clever programmers can craft their code in ways that don't require *goto*. The result is something more readable — and that's the key. Code that contains lots of *goto* statements can be difficult to follow, leading experienced programmers to describe it as *spaghetti code.* The *goto* statement encourages sloppy habits.

The only time *goto* might truly be necessary is when busting out of a nested loop. Even in this situation, an example would be contrived. So it's probably safe to say that you'll run your entire programming career and, hopefully, never have to deal with a *goto* statement in C.

» Avoiding the prototype

» Working with variables in a function

» Passing arguments to a function

» Returning values from a function

» Using *return* to leave a function

Chapter 10
Fun with Functions

When it comes to getting work done, it's a program's functions that do the heavy lifting. The C language comes with libraries full of functions, which help bolster the basics of the language, the keywords, the operators, and so on. When these C library functions fall short, you concoct your own functions.

Anatomy of a Function

The tools that are needed to craft your own functions are brief. After deciding the function's purpose, you give it a unique name, toss in some parentheses and curly brackets, and you're pretty much done. Of course, the reality is a bit more involved.

Constructing a function

All functions are dubbed with a name, which must be unique; no two functions can have the same name, nor can a function have the same name as a keyword or variable used in the code.

The name is followed by parentheses, which are then followed by a set of curly brackets. So, in its simplest construction, a function looks like this:

```
type function() { }
```

In the preceding example, `type` defines the function's data type, which is related to the value returned from the function. Options for `type` include all the standard C data types — *char*, *int*, *float*, *double* — and also *void* for cheap functions that don't return anything.

`function` is the function's name. It's followed by a pair of parentheses, which can, optionally, contain values passed to the function. These values are called *arguments*. Not every function features arguments.

Central to a function are its statements, enclosed in curly brackets. These statements are what make the function do its thing.

Functions that return a value must use the *return* keyword. The *return* statement either ends the function directly or passes a value back to the statement that called the function. For example:

```
return;
```

This statement ends a function and doesn't pass on a value. Any statements in the function after *return* are ignored.

```
return(something);
```

This statement passes the value of the `something` variable back to the statement that called the function. The `something` must be of the same data type as the function, an *int*, the *float*, and so on.

Functions that don't return values are declared of the *void* type. Those functions end with the last statement held in the curly brackets; a *return* statement isn't required.

REMEMBER

One more important thing! Functions must be *prototyped* in your code. This is so that the compiler understands the function and sees to it that you use it properly. The prototype describes the value returned and any arguments the function requires. The prototype must appear in your source code before any statement calls the function. Listing 10-1 shows a prototype example at Line 3.

LISTING 10-1: Basic Function; No Return

```
#include <stdio.h>

void prompt();      /* function prototype */

int main()
{
    int loop;
    char input[32];

    loop=0;
    while(loop<5)
    {
        prompt();
        fgets(input,32,stdin);
        loop=loop+1;
    }
    return(0);
}

/* Display prompt */

void prompt(void)
{
    printf("C:\\DOS> ");
}
```

Exercise 10-1: Use the source code from Listing 10-1 to create a new program. Build and run.

The program displays a prompt five times, allowing you to type various commands. Of course, nothing happens when you type commands, though you can program those actions later, if you like. Here's how this program works in regard to creating a function:

Line 3 lists the function prototype. It's essentially a copy of the first line of the function (from Line 22), but ending with a semicolon. It can also be written like this:

```
void prompt(void);
```

The function returns no values and requires no arguments, so its data type is *void* and *void* appears in the parentheses.

Line 13 accesses the function. The function is called as its own statement. It doesn't require any arguments or return any values, and it appears on a line by itself, as shown in the listing. When the program encounters that statement, program execution jumps up to the function. The function's statements are executed, and then control returns to the next line in the code after the function was called.

Lines 22 through 25 define the function itself. The function's data type, *void*, is specified on Line 22, followed by the function name, and then the parentheses with *void* specified because arguments aren't passed to the function.

The function's sole statement is held between curly brackets. The *prompt()* function uses the *printf()* function to output a prompt, which makes it seem like the function isn't necessary, but many examples of one-line functions can be found in lots of programs.

Exercise 10-2: Modify the source code from Listing 10-1 to add the *busy()* function, which contains the *while* loop now in the *main()* function. (Copy Lines 7 through 16 into the new function.) Have the *main()* function call the *busy()* function.

» C has no limit on what you can do in a function. Any statements you can stuff into the *main()* function can go into any function. Indeed, *main()* is simply another function in your program, albeit the program's chief function.

WARNING

» The *main()* function has arguments, so don't be tempted to edit its empty parentheses and stick the word *void* in there. In other words, this construct is wrong:

```
int main(void)
```

The *main()* function in C has two arguments. It's possible to avoid listing them when you're not going to use them, by keeping parentheses empty. Chapter 15 discusses using the *main()* function's arguments.

TECHNICAL STUFF

» Other programming languages may refer to a function as a *subroutine* or *procedure*.

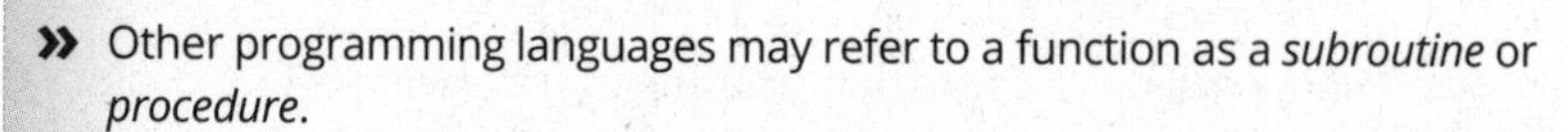

Prototyping (or not)

What happens when you don't prototype? As with anything in programming, the compiler or linker lets you know with a warning or error message when you goof — or the program just doesn't run properly. It's not the end of the world — no, not like programming a military robot or designing genetic code for a new species of Venus flytrap.

Exercise 10-3: Modify the source code from Exercise 10-1. Comment out the prototype from Line 3. Build the result.

Build errors are wonderful things, delightfully accurate yet entirely cryptic. Here are the warning and error messages generated by my compiler and linker, showing only the relevant parts of each message:

```
13 warning: implicit declaration of function 'prompt'
22 error: conflicting types for 'prompt'
```

The first warning occurs at Line 13 in my source code file, where the *prompt()* function is used inside the *main()* function. The compiler is telling you that a function is called without a prototype. As the message says, you're implicitly declaring a function. That's a no-no, but not a full-on error; the compiler generates object code, praying that the linker can resolve the unknown function.

The error occurs where the *prompt()* function dwells in the program. In my source code, it's at Line 22. The *prompt()* function's data type is incorrect because the linker assumes integer as an unknown function's data type.

You may draw the conclusion that prototyping is an absolute necessity in your C code. This assumption isn't entirely true. You can avoid prototyping by reordering the functions in your source code. As long as a function is listed before it's used, you don't need a prototype.

Exercise 10-4: Edit your source code from Exercise 10-3. Remove the function prototype that was commented out at Line 3. Cut-and-paste (move) the *prompt()* function from the bottom of the source code listing to the top, above the *main()* function. Save, build, and run.

Listing 10-2 shows what I conjured up as a solution for Exercise 10-4.

LISTING 10-2: Avoiding the Function Prototype

```
#include <stdio.h>

/* display prompt */

void prompt(void)
```

(continued)

LISTING 10-2: *(continued)*

```
{
    printf("C:\\DOS> ");
}

int main()
{
    int loop;
    char input[32];

    loop=0;
    while(loop<5)
    {
        prompt();
        fgets(input,32,stdin);
        loop=loop+1;
    }
    return(0);
}
```

In this book, as well as in my own programs, I write the *main()* function first, followed by other functions. This method is traditional in the C language, and it allows for better readability. You are free on your own to put functions first to avoid prototyping. And if you don't, keep in mind that other programmers may do it that way, so don't be surprised when you see it.

TECHNICAL STUFF

Compiler messages may feature parenthetical comments that refer to the *switch*, or traditional command-line option, that enables checking for a particular warning. For example, the warning message from Exercise 10-3 reads in full:

```
ex1003.c:13:9: warning: implicit declaration of function
   'prompt' is invalid in C99 [-Wimplicit-function-declaration]
```

The specific compiler warning switch for this error message is `-Wimplicit-function-declaration`.

Functions and Variables

I'm fond of saying that functions gotta funct. That is, they need to do something, to work as a machine that somehow manipulates input or generates output. To make your functions funct, you need to know how to employ variables to, from, and within a function.

Using variables in functions

Functions that use variables must declare those variables — just like the *main()* function does. In fact, it's pretty much the same thing. The big difference, which you must remember, is that variables declared and used within a function are local to that function. Or, to put it in the vernacular, what happens in a function stays within the function, as shown in Listing 10-3.

LISTING 10-3: Local Variables in a Function

```
#include <stdio.h>

void vegas(void);

int main()
{
    int a;

    a = 365;
    printf("In the main() function, a=%d\n",a);
    vegas();
    printf("In the main() function, a=%d\n",a);
    return(0);
}

void vegas(void)
{
    int a;

    a = -10;
    printf("In the vegas() function, a=%d\n",a);
}
```

Both the *main()* and *vegas()* functions declare and use *int* variable a. The variable is assigned the value 365 in *main()* at Line 9. In the *vegas()* function, variable a is assigned the value -10 at Line 20. Can you predict the program's output for the *printf()* function on Line 12?

Exercise 10-5: Create a new project using the source code from Listing 10-3. Build and run.

Here's the output I see:

```
In the main() function, a=365
In the vegas() function, a=-10
In the main() function, a=365
```

The same variable name is used in both functions, yet it holds a different value in each. That's because variables in C are local to their functions: One function cannot change the value of a variable in another function, even if both variables sport the same type and name.

- My admonition earlier in this book about not duplicating variable names doesn't hold for variables in other functions. You could have 16 functions in your code, and each function uses the alpha variable. That's perfectly okay. Even so:
- You need not use the same variable names in all functions. The *vegas()* function from Listing 10-3 could have declared its variable name as pip or wambooli.
- To allow multiple functions to share a variable, you specify an external or global variable. That topic is avoided until Chapter 16.

TECHNICAL STUFF

- Variables local to their functions are determined to be of the *auto* storage class. The *auto* keyword could prefix variables local to a function, as in auto int a, though it would be considered anachronistic.

Sending a value to a function

The key way to make a function funct is to give it something to chew on — some data. The process is referred to as *passing an argument to a function*, where the term *argument* is used in C programming to refer to an option or a value. It comes from the mathematical term for variables in a function, so no bickering is anticipated.

Arguments are specified in the function's parentheses. An example is the *puts()* function, which accepts a string as an argument, as in

```
puts("You probably shouldn't have chosen that option.");
```

The *fgets()* function swallows three arguments at once:

```
fgets(buffer,27,stdio);
```

Arguments can be variables, constants, or literal values, and multiple arguments are separated by commas. The number and type of values that a function requires must be specified when the function is written and for its prototype as well. Listing 10-4 illustrates an example.

LISTING 10-4: Passing a Value to a Function

```
#include <stdio.h>

void graph(int count);

int main()
{
    int value;

    value = 2;

    while(value<=64)
    {
        graph(value);
        printf("Value is %d\n",value);
        value = value * 2;
    }
    return(0);
}

void graph(int count)
{
    int x;

    for(x=0;x<count;x=x+1)
        putchar('*');
    putchar('\n');
}
```

When a function consumes an argument, you must clearly tell the compiler what type of argument is required. In Listing 10-4, both the prototype at Line 3 and the *graph()* function's definition at Line 20 state that the argument must be an *int*. The variable `count` is used as the *int* argument, which then serves as the variable's name inside the function.

The *graph()* function is called in Line 13, in the midst of the *while* loop. It's called using the `value` variable. That's okay; the variable name used in a function call need not match the variable name used inside the function. Only the variable's data type must match, and both `count` and `value` are *int* types.

The *graph()* function, from Line 20 through Line 27, displays a row of asterisks. The length of the row (in characters) is determined by the value sent to the function.

Exercise 10-6: Fire up your editor and feverishly type the source code from Listing 10-4. Save it. Build it. Can you guess what the output might look like before running?

Functions don't necessarily need to consume variables. The *graph()* function from Listing 10-4 can gobble any *int* value, including an immediate value (literal) or a constant.

Exercise 10-7: Edit the source code from Exercise 10-6, changing Line 13 so that the *graph()* function is passed a constant value of 64. Build and run.

WARNING

It's possible to pass a string to a function, but until you've read Chapter 12 on arrays and especially Chapter 18 on pointers, I don't recommend it. A string is really an array, and it requires special C language magic to pass an array to a function.

Sending multiple values to a function

C offers no limit on how many arguments a function can handle. As long as you properly declare the arguments as specific data types and separate them all with commas, you can stack 'em up like commuters on a morning train, similar to this prototype:

```
void railway(int engine, int boxcar, int caboose);
```

In the preceding line, the *railway()* function is prototyped. It requires three *int* arguments: `engine`, `boxcar`, and `caboose`. The function must be passed three arguments, as shown in the prototype.

Exercise: 10-8: Modify the source code from Listing 10-4 so that the *graph()* function accepts two arguments; the second is the character to display.

Creating functions that return values

A great majority of the C language functions return a value; that is, they generate something. Your code may not use the values, but they're returned anyway. For example, both *putchar()* and *printf()* return values, and I've rarely seen a program use these values.

Listing 10-5 illustrates a function that is sent a value and then returns another value. This is the way most functions work, though some functions return values without necessarily receiving any. For example, *getchar()* returns input but doesn't require any arguments. In Listing 10-5, the *convert()* function accepts a Fahrenheit value and returns its Celsius equivalent.

LISTING 10-5: A Function That Returns a Value

```
#include <stdio.h>

float convert(float f);

int main()
{
    float temp_f,temp_c;

    printf("Temperature in Fahrenheit: ");
    scanf("%f",&temp_f);
    temp_c = convert(temp_f);
    printf("%.1fF is %.1fC\n",temp_f,temp_c);
    return(0);
}

float convert(float f)
{
    float t;

    t = (f - 32) / 1.8;
    return(t);
}
```

Line 3 in Listing 10-5 declares the *convert()* function's prototype. The function requires a floating-point value and returns a floating-point value.

The *convert()* function is called in Line 11. Its return value is stored in variable `temp_c` on that same line. In Line 12, *printf()* displays the original value and the conversion. The `%.1f` placeholder limits floating-point output to one digit to the left of the decimal. (See Chapter 13 for a full description of the *printf()* function's placeholders.)

The *convert()* function begins at Line 16. It uses two variables: `f` contains the *float* value passed to the function, a temperature in Fahrenheit. A local variable, `t`, is used to calculate the Celsius temperature value, declared at Line 18 and assigned by the formula on Line 20.

Line 20 uses scary math to convert the `f` Fahrenheit value into the `t` Celsius value. The parentheses surrounding `f - 32` direct the compiler to perform that part of the calculation first and then divide the result by 1.8. If you omit the parentheses, 32 is divided by 1.8 first, which generates an incorrect result. See Chapter 11 for information on the order of precedence, which describes how C prefers to do long math equations.

The *return* statement at Line 21 sends the function's result back to the caller.

Exercise 10-9: Type the source code from Listing 10-5 into your editor. Build and run.

Functions that return values can have that value stored in a variable, as shown on Line 11 of Listing 10-5, or you can also use the value immediately. For example:

```
printf("%.1fF is %.1fC\n",temp_f,convert(temp_f));
```

Exercise 10-10: Edit the source code from Listing 10-5 so that the *convert()* function is used immediately in the *printf()* function. ***Hint:*** That's not the only line you need to fix up to make the change complete.

You may also notice that the *convert()* function itself has a redundant item. Do you really need the `t` variable in that function?

Exercise 10-11: Edit your solution from Exercise 10-10, this time paring out the `t` variable from the *convert()* function.

Honestly, you could simply eliminate the *convert()* function altogether because it's only one line. Still, the benefit of a function like *convert()* is that you can call it from anywhere in your code. So, rather than repeat the same thing over and over and have to edit that repeated chunk of code when something changes, you instead create a function. Such a thing is perfectly legitimate, and it's done all the time in C.

And just because I'm a good guy, but also because it's referenced earlier in this chapter, Listing 10-6 shows my final result for Exercise 10-11.

LISTING 10-6: A Tighter Version of Listing 10-5

```
#include <stdio.h>

float convert(float f);

int main()
{
    float temp_f;

    printf("Temperature in Fahrenheit: ");
    scanf("%f",&temp_f);
    printf("%.1fF is %.1fC\n",temp_f,convert(temp_f));
    return(0);
}

float convert(float f)
{
    return(f - 32) / 1.8;
}
```

The *convert()* function's math is compressed to one line, so a temporary storage variable (t from Line 18 in Listing 10-5) isn't needed.

Returning early

The *return* keyword can blast out of a function at any time, sending execution back to the statement that called the function. Or, in the case of the *main()* function, *return* exits the program. This rule holds fast even when *return* doesn't pass back a value, which is true for any *void* function. Consider Listing 10-7.

LISTING 10-7: Exiting a Function with *return*

```
#include <stdio.h>

void limit(int stop);

int main()
{
    int s;

    printf("Enter a stopping value (0-100): ");
    scanf("%d",&s);
    limit(s);
    return(0);
}

void limit(int stop)
{
    int x;

    for(x=0;x<=100;x=x+1)
    {
        printf("%d ",x);
        if(x==stop)
        {
            puts("You won!");
            return;
        }
    }
    puts("I won!");
}
```

The silly source code shown in Listing 10-7 calls a function, *limit()*, with the value read in Line 10. A loop in the function spews out numbers. If a match is made with the function's argument, a *return* statement (refer to Line 25) bails out of the function. Otherwise, execution continues and the function ends when the loop is done. No *return* function is required at the end of the function because no value is returned.

Exercise 10-12: Create a new project using the source code shown in Listing 10-7. Build and run.

Constants of the Global Kind

Constants are useful in functions because they let you change a value in only one spot and have the change reflected throughout the function. Like any function's variables, however, the *const* type of constant is valid in only one function at a time. Rather than re-create a constant for each function, you can use the `#define` preprocessor directive to create a global constant, available to all functions in the source code file.

- » Use the *const* keyword to create a constant for use within a function.
- » Refer to Chapter 6 for more details on constants and the *const* keyword.

Introducing defined constants

A *defined constant* is a shortcut — specifically, something used in a source code file to substitute for something else. This constant operates at the compiler level and affects the entire source code file. It's created by using the `#define` directive, in this format:

```
#define SHORTCUT text
```

SHORTCUT is a keyword, usually written in all caps. It's created by the compiler to represent the *text* part. The line doesn't end with a semicolon, because it's a compiler directive, not a C language statement. The defined constant you create can be used elsewhere in the code, in any function: The keyword *SHORTCUT* being replaced by the *text*.

The following line creates the defined constant `OCTO`, equal to the value 8:

```
#define OCTO 8
```

After its definition, you can use the defined constant `OCTO` anywhere in your code to represent the value 8. For example:

```
printf("Mr. Octopus has %d legs.",OCTO);
```

The preceding statement outputs this text:

```
Mr. Octopus has 8 legs.
```

The OCTO defined constant is replaced by the text 8 when the source code is compiled, which translates into the value 8.

- The #define directive is traditionally placed at the top of the source code, right after any #include directives. See the next section for an example.
- You can define strings as well:

```
#define AUTHOR "Dan Gookin"
```

  The string that's defined includes the double quotes. Anywhere the defined constant AUTHOR appears in the source code file, it's replaced by the text "Dan Gookin" — including the double quotes.

- You can even define math calculations:

```
#define CELLS 24*80
```

  The defined constant CELLS doesn't do math, but the compiler replaces its text with the expression 24*80 everywhere CELLS appears in the source code file.

- Like a variable name, a defined constant specified inside a string literal is part of the string. The preprocessor doesn't expand a defined constant when it appears in a string.
- The #define directive can be used in any source code file to create a shortcut for any text, even when the source code file has only one function.

- A defined constant is created and managed by the preprocessor. It's a search-and-replace operation; no variable "storage container" is created, no data type is specified.

Putting defined constants to use

Anytime your code uses a single value over and over — something significant, like the number of rows in a table or the maximum number of items you can stick in a shopping cart – consider using a defined constant. Because this value appears in several functions, using the #define directive makes more sense than creating multiple *const* values, the same one for each function.

Listing 10-8 shows an example where a defined constant helps in two of the code's functions.

LISTING 10-8: Preparing for Constant Updates

```
#include <stdio.h>

#define GRID 3

/* prototypes */
void forward(void);
void backwards(void);

int main()
{
    puts("Grid forward:");
    forward();
    puts("Grid backwards:");
    backwards();
    return(0);
}

void forward(void)
{
    int x,y;

    for(x=0;x<GRID;x++)
    {
        for(y=0;y<GRID;y++)
            printf("%d:%d\t",x,y);
        putchar('\n');
    }
}

void backwards(void)
{
    int x,y;

    for(x=GRID-1;x>=0;x--)
    {
        for(y=GRID-1;y>=0;y--)
            printf("%d:%d\t",x,y);
        putchar('\n');
    }
}
```

Exercise 10-13: Create a new program using the source code from Listing 10-8. Build and run.

Here is the output I see:

```
Grid forward:
0:0     0:1     0:2
1:0     1:1     1:2
2:0     2:1     2:2
Grid backwards:
2:2     2:1     2:0
1:2     1:1     1:0
0:2     0:1     0:0
```

Both functions rely upon the defined constant GRID to set their size. The *forward()* function outputs the grid from low numbers to high; the *backwards()* function does the opposite.

Note how the *for* loops are constructed in the *backwards()* function:

```
for(x=GRID-1;x>=0;x--)
```

Variable x is initialized to the value of GRID minus one, which makes its starting value equal to the ending value for the companion *for* statement in the *forward()* function. The loop continues as long as the value of variable x is greater than or equal to zero. This is how both loops generate similar values in their grid output.

Exercise 10-14: Modify the source code from Exercise 10-13 so that the grid is five items square. Build and run.

Now imagine how difficult Exercise 10-14 would be if you had to search through the code to replace each instance of 3 with 5. Imagine how ugly it would be had the value 3 appeared all over the code, not just in relation to the grid. This reason is why defined constants are used.

Exercise 10-15: Modify the source code from Listing 10-7. Add a new function, *verify()*, which confirms whether the value input is within the range from 0 to 100. The function returns the defined constant TRUE (1) if the value is within the range, or FALSE (0) if not. When a value is out of range, the program displays an error message.

3
Build Upon What You Know

IN THIS PART . . .

Discover how math works in the C language

Augment variable storage by creating arrays

Manipulate and massage text

Create structures containing variable types

Work and play at the command prompt

Explore the possibilities for using variables

Dig down deep and play with binary numbers

» Making the modulus useful

» Employing various operator shortcuts

» Working with various math functions

» Creating random numbers

» Understanding the order of precedence

Chapter 11

The Unavoidable Math Chapter

One of the reasons I shunned computers in my early life was that I feared the math. Eventually, I learned that math doesn't play a central role in programming. On one hand, you need to know *some* math, especially when a program involves complex calculations. On the other hand, it's the computer that does the math — you just punch in the formula.

In the programming universe, math is necessary but painless. Most programs involve some form of simple math. Graphics programming uses a lot of math. And games wouldn't be interesting if it weren't for random numbers. All this stuff is math. I believe that you'll find it more interesting than dreadful.

Math Operators from Beyond Infinity

Two things make math happen in C programming: math operators and math functions.

Math operators allow you to construct mathematical expressions. These are shown in Table 11-1. The math functions implement complex calculations for which symbols and characters are unavailable on the keyboard. To list all these functions in a table would occupy a lot of space.

- Chapter 5 introduces the basic math operators: +, –, *, and /. The rest aren't too heavy-duty to understand — even the oddly named modulo.
- The C language comparison operators are used for making decisions. Refer to Chapter 8 for a list.
- Logical operators are also covered in Chapter 8.
- The single equal sign (=) is an operator, but not a mathematical operator. It's the *assignment* operator, used to stuff a value into a variable.
- Bitwise operators manipulate individual bits in a value. They're covered in Chapter 17.
- Appendix C lists all the C language operators.

TABLE 11-1 C Math Operators

Operator	Function	Example
+	Addition	var=a+b
–	Subtraction	var=a-b
*	Multiplication	var=a*b
/	Division	var=a/b
%	Modulo	var=a%b
++	Increment	var++
--	Decrement	var--
+	Unary plus	+var
–	Unary minus	-var

Incrementing and decrementing

Here's a handy trick, especially for those loops in your code: the increment and decrement operators. They're insanely useful.

To add one to a variable's value, use ++, as in

```
var++;
```

After this statement is executed, the value of variable var is increased *(incremented)* by 1. It's the same as writing this code:

```
var=var+1;
```

You'll find ++ used all over, especially in *for* loops; for example:

```
for(x=0;x<100;x++)
```

This looping statement repeats 100 times. It's much cleaner than writing the alternative:

```
for(x=0;x<100;x=x+1)
```

Exercise 11-1: Code a program that outputs this phrase ten times: "Get off my lawn, you kids!" Use the incrementing operator ++ in the *for* looping statement.

Exercise 11-2: Rewrite your answer for Exercise 11-1 using a *while* loop.

The ++ operator's opposite is the decrementing operator --, which is two minus signs. This operator decreases the value of a variable by 1; for example:

```
var--;
```

The preceding statement is the same as

```
var=var-1;
```

Exercise 11-3: Write a program that displays values from -5 through 5 and then back to -5 in increments of 1. The output should look like this:

```
-5 -4 -3 -2 -1 0 1 2 3 4 5 4 3 2 1 0 -1 -2 -3 -4 -5
```

This program can be a bit tricky, so rather than have you look up my solution on the web, I'm illustrating it in Listing 11-1. Please don't look ahead until you've attempted to solve Exercise 11-3 on your own.

LISTING 11-1: Counting Up and Down

```
#include <stdio.h>

int main()
{
    int c;

    for(c=-5;c<5;c++)
        printf("%d ",c);
    for(;c>=-5;c--)
        printf("%d ",c);
    putchar('\n');
    return(0);
}
```

The crux of what I want you to see happens at Line 9 in Listing 11-1, but it also plays heavily off the first *for* statement at Line 7. You might suspect that a loop counting from -5 to 5 would have the value 5 as its stop condition, as in

```
for(c=-5;c<=5;c++)
```

The problem with this approach is that the value of c is incremented to trigger the end of the loop, which means that c equals 6 when the first *for* loop is done. If c remains less than 5, as is done at Line 7, then c is automatically set to 5 when the second loop starts. Therefore, in Line 9, no initialization of variable x in the *for* statement is necessary.

Exercise 11-4: Construct a program that displays values from -10 to 10 and then back down to -10. Step in increments of 1, as was done in Listing 11-1, but use two *while* loops to display the values.

Prefixing the ++ and -- operators

The ++ operator always increments a variable's value, and the -- operator always decrements. Knowing that, consider this statement:

```
a=b++;
```

If the value of variable b is 16, you know that its value will be 17 after the ++ operation. So what's the value of variable a — 16 or 17?

As a rule, C language math equations are read from left to right. (Refer to the later section "The Holy Order of Precedence" for specifics.) Based on this rule, after the preceding statement executes, the value of variable a is 16, and the value of variable b is 17. Right?

The source code in Listing 11-2 answers the question of what happens to variable a when you increment variable b on the right side of the equal sign (the assignment operator).

LISTING 11-2: What Comes First — the = or the ++?

```
#include <stdio.h>

int main()
{
    int a,b;

    b=16;
    printf("Before, a is unassigned and b=%d\n",b);
    a=b++;
    printf("After, a=%d and b=%d\n",a,b);
    return(0);
}
```

Exercise 11-5: Type the source code from Listing 11-2. Save, build, and run.

When you place the ++ or -- operator after a variable, it's called *post-incrementing* or *post-decrementing*, respectively. If you want to increment or decrement the variable before it's used, you place ++ or -- to the left of the variable's name. For example:

```
a=++b;
```

In the preceding line, the value of b is incremented, and then it's assigned to variable a.

Exercise 11-6: Rewrite the source code from Listing 11-2 so that the equation in Line 9 increments the value of variable b before it's assigned to variable a.

And what of this construction:

```
a=++b++;
```

Never mind! The ++var++ monster is an error.

Discovering the remainder (modulus)

Of all the basic math operator symbols, % is most likely the strangest. No, it's not the percentage operator. It's the *modulus* operator. It calculates the remainder of one number divided by another, which is a concept easier to show than to discuss.

Listing 11-3 codes a program that lists the results of modulus 5 and a bunch of other values, ranging from 0 through 29. The value 5 is a constant, `mo`, which you can easily change later.

LISTING 11-3: Displaying Modulus Values

```
#include <stdio.h>

int main()
{
    const int value = 5;
    int a;

    printf("Modulus %d:\n",value);
    for(a=0;a<30;a++)
        printf("%d %% %d = %d\n",a,value,a%value);
    return(0);
}
```

Line 11 displays the modulus results. The `%%` placeholder (in Line 10) merely displays the `%` character, so don't let it throw you.

Exercise 11-7: Type in the source code from Listing 11-3. Save, build, and run.

Now that you can see the output, I can better explain how a modulus operation works. You see its calculation as the remainder of the first value divided by the second. So `20 % 5` is 0, but `21 % 5` is 1.

Exercise 11-8: Change the `value` constant in Listing 11-3 to 3. Build and run.

Saving time with assignment operators

If you're a fan of the `++` and `--` operators (and I certainly am), you'll enjoy the operators listed in Table 11-2. They're the math assignment operators, and like the increment and decrement operators, they not only do something useful — they also look really cool and confusing in your code.

TABLE 11-2 **C Math Assignment Operators**

Operator	Function	Shortcut for	Example
+=	Addition	x=x+n	x+=n
-=	Subtraction	x=x-n	x-=n
*=	Multiplication	x=x*n	x*=n
/=	Division	x=x/n	x/=n
%=	Modulo	x=x%n	x%=n

Math assignment operators do nothing new, but they work in a special way. Quite often in C, you need to modify a variable's value. For example:

```
alpha=alpha+10;
```

This statement increases the value of variable `alpha` by 10. In C, you can write the same statement by using an assignment operator as follows:

```
alpha+=10;
```

Both versions of this statement accomplish the same thing, but the second example is punchier and more cryptic, which seems to delight many C programmers. See Listing 11-4.

LISTING 11-4: Assignment Operator Heaven

```
#include <stdio.h>

int main()
{
    float alpha;

    alpha=501;
    printf("alpha = %.1f\n",alpha);
    alpha=alpha+99;
    printf("alpha = %.1f\n",alpha);
    alpha=alpha-250;
    printf("alpha = %.1f\n",alpha);
    alpha=alpha/82;
    printf("alpha = %.1f\n",alpha);
    alpha=alpha*4.3;
    printf("alpha = %.1f\n",alpha);
    return(0);
}
```

Exercise 11-9: Type the source code from Listing 11-4 into your text editor. Change Lines 9, 11, 13, and 15 so that assignment operators are used. Build and run.

TIP

When you use the assignment operator, keep in mind that the = character comes *last*. You can easily remember this tip by swapping the operators; for example:

```
alpha=-10;
```

This statement assigns the value -10 to the variable `alpha`. But the statement

```
alpha-=10;
```

decreases the value of `alpha` by 10.

Exercise 11-10: Write a program that outputs the numbers from 5 through 100 in increments of 5.

Math Function Mania

When keyboard characters and symbols for operators run dry, the C language resorts to employing various functions for mathematical operations. Those times that you're desperate to find the arctangent of an angle, you can whip out the *atan()* function and, well, there you go.

- Most math functions require including the `math.h` header file in your code. Some functions may also require the `stdlib.h` header file, where *stdlib* means *st*andard *lib*rary.
- If you're compiling at the command prompt in Linux, you may be required to include the math library in the command that builds the program. For example:

  ```
  clang -Wall ex1111.c -lm
  ```

  The *clang* compiler is used in the preceding command with the `-Wall` switch (all warnings), followed by the source code filename, `ex1111.c`. At the end of the command, the `-l` (little L) switch is followed by `m`, the name of the math library. This switch directs the linker to add the math library, which helps it incorporate various math functions.

Exploring some common math functions

Not everyone is going to employ their C language programming skills to help pilot a rocket safely across space and into orbit around Titan. No, it's more likely that you'll attempt something far more down-to-earth. Either way, the work will most likely be done by employing math functions. I've listed some common ones in Table 11-3.

TABLE 11-3 Common, Sane Math Functions

Function	#include	What It Does
sqrt()	math.h	Calculates the square root of a floating-point value
pow()	math.h	Returns the result of a floating-point value raised to a certain power
abs()	stdlib.h	Returns the absolute value (positive value) of an integer
ceil()	math.h	Rounds up a floating-point value to the next whole number (nonfractional) value
floor()	math.h	Rounds down a floating-point value to the next whole number

All the functions listed in Table 11-3, save for the *abs()* function, deal with floating-point values. The *abs()* function works only with integers.

TIP

You can look up function references in the *man* pages, as described in Chapter 1.

Listing 11-5 is littered with a smattering of math functions from Table 11-3. The compiler enjoys seeing these functions, as long as you remember to include the `math.h` header file at Line 2.

LISTING 11-5: Math Mania Mangled

```
#include <stdio.h>
#include <math.h>

int main()
{
    float result,value;

    printf("Input a float value: ");
    scanf("%f",&value);
    result = sqrt(value);
    printf("The square root of %.2f is %.2f\n",
            value,result);
```

(continued)

LISTING 11-5: *(continued)*

```
    result = pow(value,3);
    printf("%.2f to the 3rd power is %.2f\n",
            value,result);
    result = floor(value);
    printf("The floor of %.2f is %.2f\n",
            value,result);
    result = ceil(value);
    printf("And the ceiling of %.2f is %.2f\n",
            value,result);
    return(0);
}
```

I've wrapped the long *printf()* statements in Listing 11-5 so that they fit on the page in this book. You need not wrap long statements in your own source code files.

Exercise 11-11: Type the source code from Listing 11-5 into your editor. Save. Build the project. Run it and try various values as input to peruse the results.

REMEMBER

If you see any "undefined reference" errors from the linker at the Linux command prompt, add the `-lm` switch to the command-line compiling options, as described earlier in this chapter.

Exercise 11-12: Write a program that displays the powers of 2, from 2^0 through 2^{10}. These are the Holy Numbers of Computing.

» The math functions listing in Table 11-3 are only a small sampling of the variety available.

TIP

» If your code requires some sort of mathematical operation, check the C library documentation, the *man* pages, to see whether that specific function exists.

» On a Unix system, type **man 3 math** to see a list of the C library's math functions.

TECHNICAL STUFF

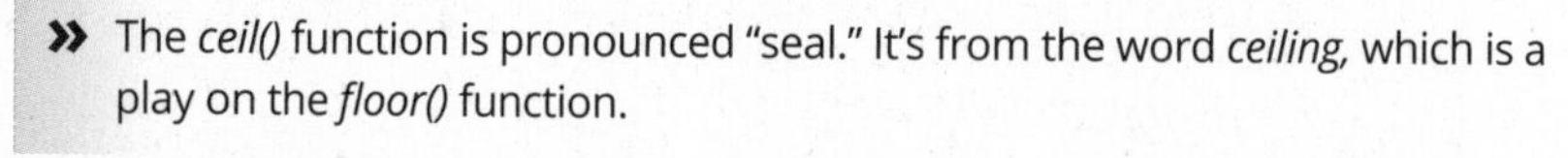

» The *ceil()* function is pronounced "seal." It's from the word *ceiling,* which is a play on the *floor()* function.

Suffering through trigonometry

I won't bother to explain trigonometry to you. If your code needs a trig function, you know why. But what you probably don't yet know is that trigonometric functions in C — and, indeed, in all programming languages — use radians, not degrees.

What's a radian?

Glad you asked. A *radian* is a measurement of a circle or, specifically, an arc. It uses the value π (pi) instead of degrees, where π is a handy circle measurement. So instead of a circle having 360 degrees, it has 2π radians. That works out to 6.2831 (which is 2×3.1415) radians in a circle. Figure 11-1 illustrates this concept.

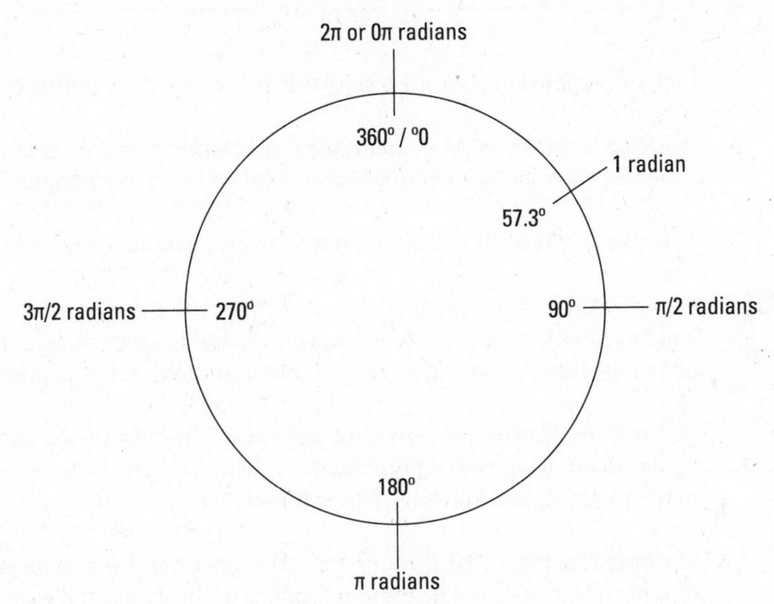

FIGURE 11-1: Degrees and radians.

For your trigonometric woes, one radian equals 57.2957795 degrees, and one degree equals 0.01745329 radians. So when you do your angle math, you need to translate between human degrees and C language radians. Consider Listing 11-6.

LISTING 11-6: Convert Degrees to Radians

```
#include <stdio.h>

int main()
{
    float degrees,radians;

    printf("Enter an angle in degrees: ");
    scanf("%f",&degrees);
    radians = 0.0174532925*degrees;
    printf("%.2f degrees is %.2f radians.\n", degrees,radians);
    return(0);
}
```

Line 10 is split in Listing 11-6 so that it fits on the page in this book.

Exercise 11-13: Type the source code from Listing 11-6 into your editor. Build and run. Test with the value 180, which should be equal to π radians (3.14).

Exercise 11-14: Write a program that converts from radians to degrees.

REMEMBER

Though C has many trigonometric functions, the three basic ones are *sin()*, *cos()*, and *tan()*, which calculate the sine, cosine, and tangent of an angle, respectively. Remember that these angles are measured in radians, not degrees.

Oh, and remember that you need the `math.h` header file to make the compiler happy about using the trig functions. And you possibly add the `-lm` switch to link in the math library. You have lots to remember.

The best programs that demonstrate trig functions are graphical in nature. This type of code would take pages to reproduce in this book, and even then I'd have to pick a platform (Windows, for example) on which the code would run. Rather than do that, I've concocted Listing 11-7 for your trigonometric enjoyment.

LISTING 11-7: Having Fun with Trigonometry

```
#include <stdio.h>
#include <math.h>

#ifndef M_PI
#define M_PI 3.14159
#endif
```

```
int main()
{
    const float amplitude=70;
    const float wavelength=0.1;
    float graph,s,x;

    for(graph=0;graph<M_PI;graph+=wavelength)
    {
        s = sin(graph);
        for(x=0;x<s*amplitude;x++)
            putchar('*');
        putchar('\n');
    }
    return(0);
}
```

The preprocessor directives in Lines 4, 5, and 6 test for the presence of the defined constant `M_PI`, which should dwell in the `math.h` header file. The `#ifndef` directive reads "if not defined." If `M_PI` isn't defined, the next line uses the `#define` directive to create it. The `#endif` directive ends the decision-making process. When complete, the code can use the `M_PI` defined constant to represent the value of π.

Exercise 11-15: Type the source code from Listing 11-7 into your editor. Before you build and run, try to guess what the output could be.

Exercise 11-16: Modify the code from Listing 11-7 so that a cosine wave is displayed. Don't get lazy on me! A cosine wave looks best when you cycle from 0 to 2π. Modify your code so that you get a good, albeit character-based, representation of the curve.

No, Exercise 11-16 isn't easy. You need to compensate for the negative cosine values when drawing the graph.

- One radian equals 57.2957795 degrees, or 360/π/2.
- One degree equals 0.0174532925 radians., or π/360*2.

It's Totally Random

One mathematical function that's relatively easy to grasp is the *rand()* function. It generates random numbers. Though such a task may seem silly, it's the basis for just about every computer game ever invented. Random numbers are a big deal in programming.

TECHNICAL STUFF

A computer cannot generate truly random numbers. Instead, it produces what are known as *pseudo-random numbers*. The reason is that conditions inside the computer can be replicated. Therefore, serious mathematicians scoff that any value a computer calls random isn't a truly random number. Can you hear them scoffing? I can.

Spewing random numbers

The *rand()* function is the simplest of C's random-number functions. It requires the `stdlib.h` header file, and it coughs up an *int* value that's supposedly random. Listing 11-8 demonstrates sample code.

LISTING 11-8: Now, That's Random

```
#include <stdio.h>
#include <stdlib.h>

int main()
{
int r,a,b;

    puts("100 Random Numbers");
    for(a=0;a<20;a++)
    {
        for(b=0;b<5;b++)
        {
            r=rand();
            printf("%d\t",r);
        }
        putchar('\n');
    }
    return(0);
}
```

Listing 11-8 uses a nested *for* loop to generate 100 random values. The *rand()* function in Line 13 generates the values. The *printf()* function in Line 14 displays the values by using the `%d` conversion character, which outputs *int* values.

Exercise 11-17: Type the source code shown in Listing 11-8 into your editor. Save, build, and run to behold 100 random values.

Exercise 11-18: Modify the code so that all values displayed are in the range 0 through 20.

TIP

Here's a hint for Exercise 11-18: Use the modulus assignment operator to limit the range of the random numbers. The format looks like this:

```
r%=n;
```

r is the number returned from the *rand()* function. `%=` is the modulus assignment operator. `n` is the range limit, plus 1. After the preceding statement, values returned are in the range 0 through *n*-1. So if you want to generate values between 1 and 100, you would use this formula:

```
value = (r % 100) + 1;
```

Making the numbers more random

Just to give some credit to the snooty mathematicians who claim that computers generate pseudo-random numbers, run the program you generated from Exercise 11-18. Observe the output. Run the program again. See anything familiar?

The *rand()* function is good at generating a slew of random values, but they're predictable values. To make the output less predictable, you need to *seed* the random-number generator. That's done by using the *srand()* function.

Like the *rand()* function, the *srand()* function requires the `stdlib.h` header, shown at Line 2 in Listing 11-9. The function requires an *unsigned int* value, `seed`, which is declared at Line 6. The *scanf()* function at Line 10 reads in the *unsigned* value by using the `%u` placeholder. Then the *srand()* function uses the `seed` value in Line 11.

LISTING 11-9: Even More Randomness

```
#include <stdio.h>
#include <stdlib.h>

int main()
{
    unsigned seed;
    int r,a,b;

    printf("Input a random number seed: ");
    scanf("%u",&seed);
    srand(seed);
    for(a=0;a<20;a++)
    {
        for(b=0;b<5;b++)
        {
            r=rand();
            printf("%d\t",r);
        }
        putchar('\n');
    }
    return(0);
}
```

The *rand()* function is used at Line 16, though the results are now based on the seed, which is set when the program runs.

Exercise 11-19: Create a new project using the source code shown in Listing 11-9. Build it. Run the program a few times, trying different seed values. The output is different every time.

Alas, the random values that are generated are still predictable when you type the same seed number. In fact, when the value 1 is used as the seed, you see the same "random" values you saw in Exercise 11-17, when you didn't even use *srand()*! And how many games have you played where you're asked to seed the randomizer? None.

There must be a better way.

The best way to write a random-number generator is not to ask the user to type a seed, but rather to fetch a seed from elsewhere. In Listing 11-10, the seed value is pulled from the system clock by using the *time()* function.

LISTING 11-10: More Truly Random than Ever

```
#include <stdio.h>
#include <stdlib.h>
#include <time.h>

int main()
{
    int r,a,b;

    srand((unsigned)time(NULL));
    for(a=0;a<20;a++)
    {
        for(b=0;b<5;b++)
        {
            r=rand();
            printf("%d\t",r);
        }
        putchar('\n');
    }
    return(0);
}
```

Chapter 21 covers programming time functions in C. Without getting too far ahead, the *time()* function returns information about the current time of day, a value that's constantly changing. The `NULL` argument helps solve some problems that I don't want to get into right now, but suffice it to say that *time()* returns a value that changes every second.

The *time()* function requires inclusion of the `time.h` header file, shown at Line 3.

The `(unsigned)` part of the statement ensures that the value returned by the *time()* function is an *unsigned* integer (not negative). This technique is known as *typecasting*, which is covered in Chapter 16.

The bottom line is that the *srand()* function is passed a seed value, courtesy of the *time()* function, and the result is that the *rand()* function generates values that are more random than you'd get otherwise.

Exercise 11-20: Type the source code from Listing 11-10 and build the project. Run it a few times to ensure that the numbers are as random as the computer can get them.

Exercise 11-21: Rewrite your solution to Exercise 8-6 (from Chapter 8) so that a random number is generated to make the guessing game more interesting but perhaps not entirely fair. Display the random number if they fail to guess it.

The Holy Order of Precedence

Before you flee the tyranny of the Unavoidable Math Chapter, you need to know about the order of precedence. It's not a religious order, and it has nothing to do with guessing the future. It's about ensuring that the math equations you code in C represent what you intend.

Getting the order correct

Consider the following puzzle. Can you guess the value of the variable `answer`?

```
answer = 5 + 4 * 3;
```

As a human, reading the puzzle from left to right, you'd probably answer 27: 5 + 4 is 9 times 3 is 27. This answer is correct — for a human. The computer, however, would answer 17.

The computer isn't wrong — it just assumes that multiplication is more important than addition. Therefore, that part of the equation gets calculated first. To the computer, the order of operations is based on which operators are used. To put it another way, multiplication has *precedence* over addition.

TIP

You can remember the basic order of precedence for the basic math operators like this:

First: Multiplication, Division

Second: Addition, Subtraction

The clever mnemonic for the basic order of precedence is, "My Dear Aunt Sally." For more detail on the order of precedence for all C language operators, see Appendix G.

Exercise 11-22: Write a program that evaluates the following equation, displaying the result:

```
20 - 5 * 2 + 42 / 6
```

See whether you can guess the output before the program runs.

Exercise 11-23: Modify the code from Exercise 11-22 so that the program evaluates the equation

```
12 / 3 / 2
```

No, that's not a date. It's 12 divided by 3 divided by 2.

Forcing order with parentheses

The order of precedence can be fooled by using parentheses. As far as the C language is concerned, anything happening within parentheses is evaluated first in any equation. So even when you forget the order of precedence, you can force it by hugging parts of an equation with parentheses.

WARNING

Math ahead!

Exercise 11-24: Code the following equation so that the result equals 14, not 2:

```
12 - 5 * 2
```

Exercise 11-25: Code the following equation (from Exercise 11-22) so that addition and subtraction take place before multiplication and division. If you do it correctly, the result is 110:

```
20 - 5 * 2 + 42 / 6
```

REMEMBER

- The code you write may deal more with variables than with literal or immediate values, so you must understand the equation and what's being evaluated. For example, if you need to add the number of full-time and part-time employees before you divide by the total payroll, put the first two values in parentheses.
- Beyond the order of precedence, parentheses add a level of readability to the code, especially in long equations. Even when parentheses aren't necessary, consider adding them if the result is more readable code.
- Appendix G lists the full order of precedence for all C language operators.

» Creating an array

» Understanding character arrays

» Sorting values in an array

» Working with multidimensional arrays

» Sending an array to a function

Chapter 12

Give Me Arrays

When I first learned to program, I avoided the topic of arrays. They didn't make sense to me. Array variables sport their own methods and madness, which is different from working with single variables in C. Rather than shun this topic and skip ahead to the next chapter (which isn't any easier), consider embracing the array as a lovely, weird, and useful tool.

Behold the Array

Humans enjoy grouping things, mostly done by type: People collect spoons, for example, or coins or husbands. These items are easily grouped because they're of similar types. The C language, too, has ways to group values of similar types, like a row of variables in a queue. The word used in C is *array*.

Avoiding arrays

At some point in your programming career, an array becomes painfully inevitable. As an example, consider Listing 12-1. The code asks for and displays your three top scores, presumably from a game.

LISTING 12-1: High Scores, the Awful Version

```
#include <stdio.h>

int main()
{
    int highscore1,highscore2,highscore3;

    printf("Your highest score: ");
    scanf("%d",&highscore1);
    printf("Your second highest score: ");
    scanf("%d",&highscore2);
    printf("Your third highest score: ");
    scanf("%d",&highscore3);

    puts("Here are your high scores");
    printf("#1 %d\n",highscore1);
    printf("#2 %d\n",highscore2);
    printf("#3 %d\n",highscore3);

    return(0);
}
```

The code in Listing 12-1 asks for three integer values. Input is stored in the three *int* variables declared in Line 5. Lines 15 through 17 output the values. Simple.

Exercise 12-1: Type the source code from Listing 12-1 into the editor. Build and run.

Typing that code can be a lot of work, right? Thank goodness it's only the top three scores. Or did I speak too soon?

Exercise 12-2: Modify the source code from Listing 12-1 so that the fourth-highest score is added. Build and run.

For the next exercise, modify the code to provide for ten high scores. No! Never mind. You can just learn about arrays instead.

Understanding arrays

An *array* is a series of variables of the same type: a dozen *int* variables, two or three *double* variables, or a string of *char* variables. The array doesn't contain all the same values. No, it's more like a series of cubbyholes into which you stick different values.

An array is declared like any other variable. It's given a data type and a name and then also a set of square brackets. The following statement declares the `highscore` array:

```
int highscore[];
```

This declaration is incomplete; the compiler doesn't yet know how many items, or *elements*, are in the array. So if the `highscore` array were to hold three elements, it would be declared like this:

```
int highscore[3];
```

This array contains three elements, each of them its own *int* value. The elements are accessed and assigned like this:

```
highscore[0] = 750;
highscore[1] = 699;
highscore[2] = 675;
```

REMEMBER

An array element is referenced by its index number in square brackets. The first item is index 0, which is something you must remember. In C, you start counting at 0, which has its advantages, so don't think it's stupid.

In the preceding example, the first array element, `highscore[0]`, is assigned the value 750; the second element, 699; and the third, 675.

After initialization, an array variable is used like any other variable in your code:

```
var = highscore[0];
```

This statement stores the value of array element `highscore[0]` to *int* variable `var`. If `highscore[0]` is equal to 750, `var` is equal to 750 after the statement executes.

Exercise 12-3: Rewrite the source code from your solution to Exercise 12-2 using an array as described in this section — but keep in mind that your array holds four values, not three.

Many solutions exist for Exercise 12-3. The brute force solution has you stuffing each array variable individually, line after line, similar to the source code in Listing 12-1. A better, more insightful solution is offered in Listing 12-2.

LISTING 12-2: High Scores, a Better Version

```
#include <stdio.h>

int main()
{
    int highscore[4];
    int x;

    for(x=0;x<4;x++)
    {
        printf("Your #%d score: ",x+1);
        scanf("%d",&highscore[x]);
    }

    puts("Here are your high scores");
    for(x=0;x<4;x++)
        printf("#%d %d\n",x+1,highscore[x]);

    return(0);
}
```

Most of the code from Listing 12-2 should be familiar to you, albeit the new array notation. The `x+1` arguments in the *printf()* statements (Lines 10 and 16) allow you to use the `x` variable in the loop but output its value starting with 1 instead of 0. Although C likes to start numbering at 0, humans still prefer starting at 1.

Exercise 12-4: Type the source code from Listing 12-2 into your editor and build a new program. Run it.

Though the program's output is pretty much the same as the output in Exercises 12-2 and 12-3, the method is far more efficient, as proven by working Exercise 12-5:

Exercise 12-5: Modify the source code from Listing 12-2 so that the top ten scores are input and displayed. (You knew this one was coming.)

Imagine how you'd have to code the answer to Exercise 12-5 if you chose not to use arrays!

REMEMBER

- The first element of an array is 0.
- When declaring an array, use the full number of elements, such as 10 for ten elements. Even though the elements are indexed from 0 through 9, you still must specify 10 when declaring the array's size.

Initializing an array

As with any variable in C, you can initialize an array when it's declared. The initialization requires a special format, similar to this statement:

```
int highscore[] = { 750, 699, 675 };
```

The number in the square brackets isn't necessary when you initialize an array, as shown in the preceding example. The reason is that the compiler is smart enough to count the elements and allocate the array's storage automatically.

Exercise 12-6: Write a program that displays the stock market closing numbers for the past five days. Use an initialized array, `marketclose[]`, to hold the values. The output should look something like this:

```
Stock Market Close
Day 1: 24164.95
Day 2: 24107.08
Day 3: 24643.63
Day 4: 24400.93
Day 5: 23728.53
```

Exercise 12-7: Write a program that uses two arrays. The first array is initialized to the values 10, 12, 14, 15, 16, 18, and 20. The second array is the same size but not initialized. In the code, fill the second array with the square root of each of the values from the first array. Output the results.

Playing with character arrays (strings)

You can create an array using any of the C language's data types. A *char* array, however, is a little different when it's a string.

As with any array, you can declare a *char* array initialized or not. The format for an initialized *char* array can look like this:

```
char letters[] = { 'c', 'a', 't' };
```

This array declaration is for a 3-element character array, containing the letters `c`, `a`, and `t`. This example is a character array, not a string.

```
char pet[] = "cat";
```

The preceding array declaration creates a string. The initialization text is enclosed in double quotes. Further, because a string literal is used, the compiler automatically adds the terminating null character: \0. This addition is what makes the array a string and not a collection of characters — an important concept to understand.

Here is the same string declaration written as individual characters:

```
char cat[] = { 'c', 'a', 't', '\0' };
```

Each array element in the preceding line is defined as its own *char* value, including the \0 character that terminates the string. (I believe that you'll find the double-quote method far more effective at declaring strings.)

The code in Listing 12-3 plods through the *char* array one character at a time. The `index` variable is used as, well, the index. The *while* loop spins until the \0 character at the end of the string is encountered. A final *putchar()* function (in Line 14) kicks in a newline.

LISTING 12-3: Displaying a *char* Array

```
#include <stdio.h>

int main()
{
    char sentence[] = "Random text";
    int index;

    index = 0;
    while(sentence[index] != '\0')
    {
        putchar(sentence[index]);
        index++;
    }
    putchar('\n');
    return(0);
}
```

Exercise 12-8: Type the source code from Listing 12-3 into your editor. Build and run the program.

The *while* loop in Listing 12-3 works like many string display routines found in the C library. These functions probably use pointers instead of arrays, which is a topic unleashed in Chapter 18. Beyond that bit o' trivia, you could replace Lines 8 through 14 in the code with the line

```
puts(sentence);
```

or even with this one:

```
printf("%s\n",sentence);
```

WARNING

When the *char* array is used as a function's argument, as shown in the preceding line, the square brackets aren't necessary. If they're included, the compiler believes that you screwed up.

Working with empty *char* arrays

Just as you can declare an empty, or uninitialized, *float* or *int* array, you can create an empty *char* array. You must be precise, however: The array's size must be one character greater than the maximum length of the string to account for the terminating null character. Also, you have to ensure that whatever input fills the array doesn't exceed the array's size.

In Listing 12-4, the *char* array `firstname` at Line 5 can hold 15 characters, plus 1 for the `\0` at the end of the string. This 15-character limitation is an assumption made by the programmer; most first names are fewer than 15 characters long.

LISTING 12-4: Filling a *char* Array

```
#include <stdio.h>

int main()
{
    char firstname[16];

    printf("What is your name? ");
    fgets(firstname,16,stdin);
    printf("Pleased to meet you, %s\n",firstname);
    return(0);
}
```

An *fgets()* function in Line 8 reads in data for the `firstname` string. The maximum input size is set to 16 characters, which already accounts for the null character because *fgets()* is smart that way. The text is read from `stdin` (the function's third argument), or standard input.

Exercise 12-9: Create a new program using the source code from Listing 12-4. Build and run, using your first name as input.

Try running the program again, but fill up the buffer: Type more than 15 characters. You'll see that only the first 15 characters are stored in the array. Even the Enter key press isn't stored, which it would be otherwise when input is fewer than 15 characters.

Exercise 12-10: Modify your source code from Exercise 12-9 so that the program also asks for your last name, storing this data in another array. The program should then greet you by using both your first and last names.

Yes, the Enter key press is stored as part of your name when you type fewer than 15 characters. This effect is how input is read by the *fgets()* function. If your first name is *Dan*, the array looks like this:

```
firstname[0] == 'D'
firstname[1] == 'a'
firstname[2] == 'n'
firstname[3] == '\n'
firstname[4] == '\0'
```

The newline lurks in the string because input in C is stream oriented and Enter is part of the input stream as far as the *fgets()* function is concerned. You can fix this issue by obeying Exercise 12-11.

Exercise 12-11: Rewrite your source code from Exercise 12-10 so that the *scanf()* function is used to read in the first and last name strings.

Of course, the problem with the *scanf()* function is that it doesn't check to ensure that input is limited to 15 characters — that is, unless you direct it to do so:

Exercise 12-12: Modify the *scanf()* functions in your source code from Exercise 12-11 so that the conversion character used is written as `%15s`. Build and run.

The `%15s` conversion character tells the first *scanf()* function to read only the first 15 characters of input and place it into the *char* array (string). Any extra text is then read by the second *scanf()* function, and any extra text after that is discarded.

It's critical that you understand stream input when it comes to reading text in C. Chapter 13 offers additional information on this important topic.

Sorting arrays

Computers are designed to quickly and merrily accomplish boring tasks, such as sorting an array. In fact, they love doing it so much that "the sort" is a basic computer concept upon which many theories and algorithms have been written. It's a real snoozer topic if you're not a Mentat or a native of the planet Vulcan.

The simplest sort is the *bubble sort*, which not only is easy to explain and understand but also has a fun name. It also best shows the basic array sorting philosophy, which is to swap values between two elements.

Suppose that you're sorting an array so that the smallest values are listed first. If `array[2]` contains the value 20, and `array[3]` contains the value 5, these two elements would need to swap values. To make it happen, you use a temporary variable in a series of statements that looks like this:

```
temp=array[2];        /* Save 20 in temp */
array[2]=array[3];    /* Store 5 in array[2] */
array[3]=temp;        /* Put 20 in array[3] */
```

In a bubble sort, each array element is compared with every other array element in an organized sequence. When one value is larger (or smaller) than another, the values are swapped. Otherwise, the comparison continues, plodding through every possible permutation of comparisons in the array. Listing 12-5 demonstrates.

LISTING 12-5: A Bubble Sort

```
#include <stdio.h>

int main()
{
    const int size = 6;
    int bubble[] = { 95, 60, 6, 87, 50, 24 };
    int inner,outer,temp,x;

    /* Display original array */
    puts("Original Array:");
    for(x=0;x<size;x++)
        printf("%d\t",bubble[x]);
    putchar('\n');

    /* Bubble sort */
    for(outer=0;outer<size-1;outer++)
```

(continued)

LISTING 12-5: *(continued)*

```
    {
        for(inner=outer+1;inner<size;inner++)
        {
            if(bubble[outer] > bubble[inner])
            {
                temp=bubble[outer];
                bubble[outer] = bubble[inner];
                bubble[inner] = temp;
            }
        }
    }

    /* Display sorted array */
    puts("Sorted Array:");
    for(x=0;x<size;x++)
        printf("%d\t",bubble[x]);
    putchar('\n');

    return(0);
}
```

Listing 12-5 is long, but it's easily split into three parts, each headed by a comment:

- Lines 9 through 13 output the original array.
- Lines 15 through 27 sort the array.
- Lines 29 through 33 output the sorted array (duplicating Lines 9 through 13).

The constant `size` is declared at Line 5. This declaration allows you to easily change the array size in case you reuse this code again later (and you will).

The sort itself involves nested *for* loops: an outer loop and an inner loop. The outer loop marches through the entire array, one step at a time. The inner loop takes its position one element higher in the array and swoops through each value individually.

Exercise 12-13: Copy the source code from Listing 12-5 into your editor and create a new project, ex1213. Build and run.

Exercise 12-14: Using the source code from Listing 12-5 as a starting point, create a program that generates 40 random numbers in the range from 1 through 100 and stores those values in an array. Display that array. Sort that array. Display the results.

CHANGE AN ARRAY'S SIZE, WILL YOU?

When an array is declared in C, its size is set. After the program runs, you can neither add nor remove more elements. So if you code an array with 10 elements, as in

```
int topten[10];
```

you cannot add an 11th element to the array. Doing so leads to all sorts of woe and misery.

To use nerdy lingo, an array in C is not *dynamic*: It cannot change size after the size has been established. Other programming languages let you resize, or *redimension*, arrays, but not C. A workaround for dynamic storage does exist in C, but it involves using pointers to allocate storage. This scary topic is avoided until Chapter 18.

Exercise 12-15: Modify the source code from Exercise 12-14 so that the numbers are sorted in reverse order, from largest to smallest.

Exercise 12-16: Write a program that sorts the text in the 21-character string "C Programming is fun!"

Multidimensional Arrays

The arrays described in the first part of this chapter are known as *single-dimension* arrays: They're basically a series of values, one after the other. This organization is fine for describing items that march single file. When you need to describe items in the second or third dimension, you conjure forth a multidimensional type of array.

Making a two-dimensional array

It helps to think of a two-dimensional array as a grid of rows and columns. An example of this type of array is a chess board — a grid of eight rows and eight columns. Though you can declare a single 64-element array to handle the job of representing a chess board, a two-dimensional array works better. Such a thing is declared this way:

```
int chess[8][8];
```

The two square brackets define two different dimensions of the chess array: eight rows and eight columns. The square located at the first row and column would be referenced as chess[0][0]. The last square on that row would be chess[0][7], and the last square on the board would be chess[7][7].

In Listing 12-6, a simple tic-tac-toe board is created using a two-dimensional matrix: 3-by-3. Lines 9 through 11 fill in the matrix. Line 12 adds an X character in the center square.

LISTING 12-6: Tic-Tac-Toe

```
#include <stdio.h>

int main()
{
    char tictactoe[3][3];
    int x,y;

/* initialize matrix */
    for(x=0;x<3;x++)
        for(y=0;y<3;y++)
            tictactoe[x][y]='.';
    tictactoe[1][1] = 'X';

/* display game board */
    puts("Ready to play Tic-Tac-Toe?");
    for(x=0;x<3;x++)
    {
        for(y=0;y<3;y++)
            printf("%c ",tictactoe[x][y]);
        putchar('\n');
    }
    return(0);
}
```

Lines 14 through 21 output the matrix. As with its creation, the matrix is output by using a nested *for* loop.

Exercise 12-17: Create a new project using the source code shown in Listing 12-6. Build and run.

A type of two-dimensional array that's pretty easy to understand is an array of strings, as shown in Listing 12-7.

LISTING 12-7: An Array of Strings

```
#include <stdio.h>

int main()
{
    int const size = 3;
    char caesar[size][9] = {
        "Julius",
        "Augustus",
        "Nero"
    };
    int x,index;

    for(x=0;x<size;x++)
    {
        index = 0;
        while(caesar[x][index] != '\0')
        {
            putchar(caesar[x][index]);
            index++;
        }
        putchar('\n');
    }
    return(0);
}
```

Line 6 in Listing 12-7 declares a two-dimensional *char* array: `caesar`. The first value in square brackets is the number of items (strings) in the array, set by the constant `size`. The second value in square brackets is the maximum size required to hold the largest string. The largest string is `Augustus` with eight letters, so nine characters are required for storage, which includes the terminating `\0` or null character.

Because all items in the array's second dimension must have the same number of elements, all strings are stored using nine characters. Yep, this configuration is wasteful, but it's the way the system works. Figure 12-1 illustrates this concept.

Exercise 12-18: Type the source code from Listing 12-7 into your editor; build and run the program.

9 elements

	0	1	2	3	4	5	6	7	8
caesar[0]	J	u	l	i	u	s	\0		
caesar[1]	A	u	g	u	s	t	u	s	\0
caesar[2]	N	e	r	o	\0				

FIGURE 12-1: Storing strings in a two-dimensional array.

Lines 15 through 21 in Listing 12-7 are inspired by Exercise 12-8, earlier in this chapter. The statements basically plod through the `caesar` array's second dimension, spitting out one character at a time.

Exercise 12-19: Replace Lines 15 through 21 in Listing 12-7 with a single *puts()* function to display the string. Here's how that statement looks:

```
puts(caesar[x]);
```

When working with strings in a two-dimensional array, the string is referenced by the first dimension only.

Exercise 12-20: Modify your source code from Exercise 12-19 so that three more emperors are added to the array: Tiberius, Caligula, and Claudius.

Going crazy with three-dimensional arrays

Two-dimensional arrays are pretty common in the programming realm. Multidimensional is insane!

Well, maybe not. Three- and four-dimensional arrays have their place. The big deal is that your human brain has trouble keeping up with the various possible dimensions.

Listing 12-8 illustrates code that works with a three-dimensional array. The declaration is found at Line 5. The third dimension is simply the third set of square brackets, which effectively creates a 3D tic-tac-toe game board.

LISTING 12-8: Going 3D

```
#include <stdio.h>

int main()
{
    char tictactoe[3][3][3];
    int x,y,z;

/* initialize matrix */
    for(x=0;x<3;x++)
        for(y=0;y<3;y++)
            for(z=0;z<3;z++)
                tictactoe[x][y][z]='.';
    tictactoe[1][1][1] = 'X';

/* display game board */
    puts("Ready to play 3D Tic-Tac-Toe?");
    for(z=0;z<3;z++)
    {
        printf("Level %d\n",z+1);
        for(x=0;x<3;x++)
        {
            for(y=0;y<3;y++)
                printf("%c ",tictactoe[x][y][z]);
            putchar('\n');
        }
    }
    return(0);
}
```

Lines 8 through 12 fill the array with data, using variables `x`, `y`, and `z` as the three-dimensional coordinates. Line 13 places character X in the center cube, which gives you an idea of how individual elements are referenced.

The rest of the code from Lines 15 through 26 displays the matrix.

Exercise 12-21: Create a three-dimensional array program using the source code from Listing 12-8. Build and run.

Lamentably, the output is two-dimensional. If you'd like to code a third dimension, I'll leave that up to you.

Declaring an initialized multidimensional array

The dark secret of multidimensional arrays is that they don't really exist. Internally, the compiler sees things as single dimensions — just a long array full of elements. The double (or triple) bracket notation is used to calculate the proper offset in the array at compile time. That's okay because the compiler does the work.

You can see how multidimensional arrays translate into regular old boring arrays when you declare them already initialized. For example:

```
int grid[3][4] = {
    5, 4, 4, 5,
    4, 4, 5, 4,
    4, 5, 4, 5
    };
```

The grid array consists of three rows of four items each. As just shown, it's declared as a grid and it looks like a grid. Such a declaration works, as long as the last element doesn't have a comma after it. In fact, you can write the whole thing like this:

```
int grid[3][4] = { 5, 4, 4, 5, 4, 4, 5, 4, 4, 5, 4, 5 };
```

This statement still defines a multidimensional array, but you can see how it's really just a single-dimension array with dual indexes. In fact, the compiler is smart enough to figure out the dimensions even when you give only one of them, as in this example:

```
int grid[][4] = { 5, 4, 4, 5, 4, 4, 5, 4, 4, 5, 4, 5 };
```

In the preceding line, the compiler sees the 12 elements in an array grid, so it automatically knows that it's a 3-by-4 matrix based on the 4 in the brackets. Or you can do this:

```
int grid[][6] = { 5, 4, 4, 5, 4, 4, 5, 4, 4, 5, 4, 5 };
```

In this example, the compiler would figure that you have two rows of six elements. But the following example is just wrong:

```
int grid[][] = { 5, 4, 4, 5, 4, 4, 5, 4, 4, 5, 4, 5 };
```

The compiler isn't going to get cute. In the preceding line, it sees an improperly declared single-dimension array. The extra square brackets aren't needed.

Exercise 12-22: Rewrite the code from Exercise 12-17 so that the tic-tac-toe game board is initialized when the array is declared — including putting the X in the proper spot.

Arrays and Functions

Creating an array for use inside a function works just like creating an array for use inside the *main()* function: The array is declared, it's initialized, and its elements are used. You can also pass arrays to, and return them from, functions.

Passing an array to a function

Sending an array off to a function is pretty straightforward. The function must be prototyped with the array specified as one of the arguments. It looks like this:

```
void whatever(int nums[]);
```

This statement prototypes the *whatever()* function. The function accepts an integer array as an argument, using `nums` to represent the array within the function. The entire array — every element — is passed to the function, where it's available for fun and frolic.

When you call a function with an array as an argument, you must omit the square brackets:

```
whatever(values);
```

In the preceding line, the *whatever()* function is called using the array `values` as an argument. If you keep the square brackets, the compiler assumes that you meant only to pass a single element and that you forgot to specify which one. So this is good:

```
whatever(values);
```

But this is not good:

```
whatever(values[]);
```

And this statement passes a single element to the *nice()* function:

```
nice(value[x]);
```

The topic in this section is passing entire arrays, and the code shown in Listing 12-9 features the *showarray()* function. It's a *void* function, so it doesn't return any values, but it can manipulate the passed array.

LISTING 12-9: Mr. Function, Meet Mr. Array

```
include <stdio.h>

void showarray(int array[]);

int main()
{
    int n[] = { 2, 3, 5, 7, 11 };

    puts("Here's your array:");
    showarray(n);
    return(0);
}

void showarray(int array[])
{
    int x;

    for(x=0;x<5;x++)
        printf("%d ",array[x]);
    putchar('\n');
}
```

REMEMBER

The *showarray()* function is called at Line 10. See how the `n` array is passed without its angle brackets? Remember this format!

At Line 14, the *showarray()* function is declared with the array specified using square brackets, just like the prototype at Line 3. Within the function, the array is accessed just like it would be in the *main()* function, which you can see at Line 19.

Exercise 12-23: Type the source code from Listing 12-9 into your editor. Build and run the program to ensure that it works.

Exercise 12-24: Add a second function, *arrayinc()*, to your solution for Exercise 12-23. Make it a *void* function. The function takes an array as its argument. The function adds 1 to each value in the array. Have the *main()* function call *arrayinc()* with array `n` as its argument. Then call the *showarray()* function a second time to display the modified values in the array.

The *arrayinc()* function need not return the modified array (a topic referenced in the next section). Because of the interesting nature of arrays in C, the element's values manipulated in the one function affect the array throughout the code.

Returning an array from a function

In addition to being passed an array, a function in C can return an array. The problem is that arrays are returned only as pointers. This topic is covered in Chapter 19. But it's not the worst part:

In Chapter 19, you discover the scandalous truth that C has no arrays — that they are merely cleverly disguised pointers. (Sorry to save that revelation for the end of this chapter.) Array notation does have its place, but pointers are where the action is.

» Converting text characters

» Manipulating strings

» Working with conversion characters

» Adjusting text output

» Understanding stream input

Chapter 13
Fun with Text

A *string* is a chunk of text. It's a programming concept but also a basic part of all human communications. What a string isn't, however, is a variable type in the C language; nope, a string is an array of *char* variables. This fact doesn't make it less important. A lot of programming involves presenting text and manipulating strings. So, despite its not being invited to the official C language data type club, the string has lots of clout when it comes to writing programs.

Character Manipulation Functions

At the heart of any string of text is the *char* variable. It's a unique cubby hole, into which you stuff an integer from 0 through 255. This value is represented visually as a character — a symbol, squiggle, or whatsis, and the beloved alphabet you've been familiar with even before you learned to read.

Introducing the CTYPEs

The C language features a bevy of functions designed to test or manipulate individual characters. The functions are all defined in the `ctype.h` header file. Most programmers therefore refer to the functions as the *CTYPE functions*, where CTYPE is pronounced "see-type," and not "stoor-ye," which how a native Russian would read it.

To use the CTYPE functions, the `ctype.h` header file must be included in your source code:

```
#include <ctype.h>
```

I classify the CTYPE functions into two categories: is-testing and to-manipulation. Some of my favorite testing functions are shown in Table 13-1; manipulation functions, in Table 13-2.

TABLE 13-1 **CTYPE Testing Functions**

Function	Returns TRUE When *ch* Is
isalnum(ch)	A letter of the alphabet (upper- or lowercase) or a number
isalpha(ch)	An upper- or lowercase letter of the alphabet
iscntrl(ch)	A control code character, values 0 through 31 and 127
isdigit(ch)	A character 0 through 9
isgraph(ch)	Any printable character except for the space
islower(ch)	A lowercase letter of the alphabet, *a* to *z*
isprint(ch)	Any character that can be displayed, including the space
ispunct(ch)	A punctuation symbol
isspace(ch)	A whitespace character, space, tab, form feed, or Enter, for example
isupper(ch)	An uppercase letter of the alphabet, *A* to *Z*
isxdigit(ch)	Any hexadecimal digit, 0 through 9 or *A* through *F* (upper- or lowercase)

TABLE 13-2 **CTYPE Manipulation Functions**

Function	Returns
tolower(*ch*)	The lowercase of character *ch*
toupper(*ch*)	The uppercase of character *ch*

CTYPE functions are consistent: Testing functions begin with *is*, and manipulation functions begin with *to*.

Every CTYPE function accepts an *int* value as the argument, represented by the variable *ch* in Tables 13-1 and 13-2. These are not *char* functions!

Every CTYPE function returns an *int* value. The functions in Table 13-1 return logical TRUE or FALSE values; FALSE is 0, and TRUE is a non-zero value.

CTYPE functions are not real functions. No, they're macros defined in the `ctype.h` header file. Regardless, they look like functions and are used that way. (I write this note to prevent college sophomores from emailing me such corrections.)

Testing characters

The CTYPE functions come in most handy when testing input, determining that the proper information was typed, or pulling required information out of junk. The code in Listing 13-1 illustrates how a program can scan text, pluck out certain attributes, and then output a summary of that information.

LISTING 13-1: Text Statistics

```
#include <stdio.h>
#include <ctype.h>

int main()
{
    char phrase[] = "When in the Course of human events, it becomes
           necessary for one people to dissolve the political bands
           which have connected them with another, and to assume
           among the powers of the earth, the separate and equal
           station to which the Laws of Nature and of Nature's God
           entitle them, a decent respect to the opinions of mankind
           requires that they should declare the causes which impel
           them to the separation.";

    int index,alpha,space,punct;

    alpha = space = punct = 0;

        /* gather data */
    index = 0;
    while(phrase[index])
    {
        if(isalpha(phrase[index]))
            alpha++;
        if(isspace(phrase[index]))
            space++;
```

(continued)

LISTING 13-1: *(continued)*

```
        if(ispunct(phrase[index]))
            punct++;
        index++;
    }

        /* print results */
    printf("\"%s\"\n",phrase);
    puts("Statistics:");
    printf("%d alphabetic characters\n",alpha);
    printf("%d spaces\n",space);
    printf("%d punctuation symbols\n",punct);

    return(0);
}}
```

Listing 13-1 may seem long, but it's not; the `phrase[]` string declared at Line 6 can be anything you like — any text, a poem, or a filthy limerick. It should be long enough to have a smattering of interesting characters. Note that though the text wraps and indents in this text, you should just type one long line of text in your code.

This code also does something not yet presented in this book. I call it a *gang initialization*:

```
alpha = space = punct = 0;
```

Because each of these variables must be set to 0, you use multiple assignment operators on the same line and accomplish the task in one fell swoop.

The meat of the program's operation takes place starting after the `gather data` comment. A *while* loop steps through each character in the string. The condition for the *while* loop is `phrase[index]`. This evaluation is TRUE for each character in the array except for the last one, the null character, which evaluates to FALSE and stops the loop.

CTYPE functions are used in *if* statements as each character is evaluated at Lines 16, 18, and 20. I don't use *if-else* tests because every character must be checked. When a positive or TRUE match is found, a counter variable is incremented.

Exercise 13-1: Type the source code from Listing 13-1 into your editor. Build and run.

Exercise 13-2: Modify the source code from Listing 13-1 so that tests are also made for counting upper- and lowercase letters. Display those results as well.

Exercise 13-3: Add code to your solution to Exercise 13-2 so that a final tally of all characters in the text (the text's length) is displayed as the final statistic.

TECHNICAL STUFF

Many compilers add other is-CTYPE functions to their libraries. These functions are useful, such as *isblank()* to test for a space or tab, but they might not be available to every C compiler.

Changing characters

The CTYPE functions that begin with *to* are used to manipulate characters. The most common of these functions (the only two in the standard C library) are *toupper()* and *tolower()*, which come in handy when testing input. As an example, consider the typical yorn problem, illustrated in Listing 13-2.

LISTING 13-2: A yorn Problem

```
#include <stdio.h>
#include <ctype.h>

int main()
{
    char answer;

    printf("Would you like to blow up the moon? ");
    scanf("%c",&answer);
    answer = toupper(answer);
    if(answer=='Y')
        puts("BOOM!");
    else
        puts("The moon is safe");
    return(0);
}
```

Yorn is programmer-speak for a yes-or-no situation: The user is asked to type Y for Yes or N for No. Does the person have to type Y or y? Or can they type N or n, or would any non-Y key be considered No?

In Listing 13-2, Line 10 uses *toupper()* to convert the character input to uppercase. This way, only a single *if* condition is required in order to test for Y or y input.

Exercise 13-4: Create a new program using the source code shown in Listing 13-2. Build and run.

Exercise 13-5: Modify the source code from Listing 13-2 so that text is displayed when the user types neither Y nor N.

Exercise 13-6: Write a program that changes all uppercase letters in a string of text to lowercase and changes all lowercase letters to uppercase. Output the results.

Here's sample output from my solution to Exercise 13-6:

```
Original: ThiS Is a RANsom NoTE!
Modified: tHIs iS A ranSOM nOte!
```

Some C libraries add the *toascii()* function to the CTYPE parade. This function converts non-ASCII characters, those with codes greater than 127, to ASCII codes from 0 to 127. The *toascii()* function, however, is nonstandard.

String Functions Galore

Despite its not-a-variable type classification, the C library doesn't skimp on functions that manipulate strings. Just about anything you desire to do with a string can be done by using some of the many string functions. And when these functions fall short, you can write your own.

Reviewing string functions

Table 13-3 lists some of the C language library functions that manipulate or abuse strings.

More string functions are available than are shown in Table 13-3. Many of them do specific things that require a deeper understanding of C. The ones shown in the table are the most common found in the standard library.

TABLE 13-3 String Functions

Function	What It Does
strcat()	Appends one string to another, creating a single string out of two.
strncat()	Appends a given number of characters from one string to the end of another.
strchr()	Searches for a character within a string. The function returns that character's position from the start of the string as a pointer.
strcmp()	Compares two strings in a case-sensitive way. If the strings match, the function returns `0`.
strncmp()	Compares the first *n* characters of two strings, returning 0 if the given number of characters match.
strcpy()	Copies (duplicates) one string to another.
strncpy()	Copies a specific number of characters from one string to another.
strlen()	Returns the length of a string, not counting the `\0` or `NULL` character at the end of the string.
strrchr()	Searches for a character within a string, but in reverse. The function returns the character's position from the end of the string as a pointer.
strstr()	Searches for one string inside another string. The function returns a pointer to the string's location if it's found.

All the string functions in Table 13-3 require the `string.h` header file to be included with your source code:

```
#include <string.h>
```

TIP

On a Unix system, you can review all the string functions by typing the command **man string** in a terminal window.

WARNING

Be aware of nonstandard string functions. For example, the *strcasecmp()* function compares strings regardless of text case, though this function isn't part of the standard C library. If you use such a function in your code, it may not compile on another computer system.

Comparing text

Strings are compared by using the *strcmp()* function and its cousin *strncmp()*.

The string comparison functions return an *int* value based on the result of the comparison: 0 for when the strings are equal, or a higher or lower *int* value based on whether the first string's value is greater than (higher in the alphabet) or less than (lower in the alphabet) the second string. Most of the time, you just check for 0.

Listing 13-3 uses the *strcmp()* function in Line 13 to compare the initialized string `password` with whatever text is read at Line 11, which is stored in the `input` array. The result of this operation is stored in the *match* variable, which is used in an *if-else* decision tree at Line 14 to display the results.

LISTING 13-3: Let Me In

```
#include <stdio.h>
#include <string.h>

int main()
{
    char password[]="taco";
    char input[15];
    int match;

    printf("Password: ");
    scanf("%s",input);

    match=strcmp(input,password);
    if(match==0)
        puts("Password accepted");
    else
        puts("Invalid password. Alert the authorities.");

    return(0);
}
```

Exercise 13-7: Type the source code from Listing 13-3 into your editor. Try out the program a few times to ensure that it accepts only `taco` as the proper password.

Exercise 13-8: Eliminate the *match* variable from your code in Exercise 13-7 and use the *strcmp()* function directly in the *if* comparison. This is the way most programmers do it.

Building strings

The glue that sticks one string onto the end of another is the *strcat()* function. The term *cat* is short for con*cat*enate, which means to link together. Here's how it works:

```
strcat(first,second);
```

After this statement executes, the text from the second string is appended to the first string. Or you can use immediate values:

```
strcat(gerund,"ing");
```

This statement tacks the text ing onto the end of the gerund text array.

It's important that the destination string buffer be large enough to accommodate all the text. The compiler doesn't check for an overflow, which is a security risk. No, it's up to you, the programmer, to confirm that the target *char* array is large enough.

The code in Listing 13-4 declares two *char* arrays to hold text. Array first is twice as large as array last because it's the location where the second string's content is copied. The copying takes place at Line 13 with the *strcat()* function.

LISTING 13-4: Introductions

```
#include <stdio.h>
#include <string.h>

int main()
{
    char first[40];
    char last[20];

    printf("What is your first name? ");
    scanf("%s",first);
    printf("What is your last name? ");
    scanf("%s",last);
    strcat(first,last);
    printf("Pleased to meet you, %s!\n",first);
    return(0);
}
```

Exercise 13-9: Create a new program by using the source code from Listing 13-4. Run the program.

Exercise 13-10: Modify your source code so that a single space is concatenated to the first string before the last string is concatenated.

Though I'm careful in Listing 13-4 to avoid a buffer overflow, the *scanf()* function is used in this code. This function is weak for gathering text input. The *fgets()* function would be better, but it retains the newline in the string, which creates more problems for generating clean output.

Fun with *printf()* Formatting

The most popular output function in C has to be *printf()*. It's everyone's favorite. It's one of the first functions you learn in C. And as one of the most complex, it's one of the functions that no one ever fully knows.

The power in *printf()* lies in its formatting string. This string can be packed with plain text, escape sequences, and conversion characters, which are the little percent goobers that insert values into the text output. It's the conversion characters, or placeholders, that give *printf()* its real power, and they're also one of the function's least understood aspects.

- The *printf()* function is so popular that just about every current, trendy programming language features its own version.
- All conversion characters are listed in Appendix F.

Formatting floating point

You can use more than the basic `%f` conversion character to format floating-point values. In fact, here's the format I typically use in the *printf()* function's formatting text:

```
%w.pf
```

The *w* sets the maximum width of the entire number, including the decimal place. The *p* sets the number of characters after the decimal. For example:

```
printf("%9.2f",12.45);
```

This statement outputs four spaces and then `12.45`. These four spaces plus `12.45` (five characters total) equal the 9 in the width. Only two values are shown to the right of the decimal because `.2` is used in the `%f` conversion character.

It's possible to specify the decimal value without setting a width, but it must be prefixed by the period, as in `%.2f` (percent point-two F). Listing 13-5 shows a variety of options.

LISTING 13-5: The *printf()* Floating-Point Formatting Gamut

```
#include <stdio.h>

int main()
{
    float sample1 = 34.5;
    float sample2 = 12.3456789;

    printf("%%9.1f = %9.1f\n",sample1);
    printf("%%8.1f = %8.1f\n",sample1);
    printf("%%7.1f = %7.1f\n",sample1);
    printf("%%6.1f = %6.1f\n",sample1);
    printf("%%5.1f = %5.1f\n",sample1);
    printf("%%4.1f = %4.1f\n",sample1);
    printf("%%3.1f = %3.1f\n",sample1);
    printf("%%2.1f = %2.1f\n",sample1);
    printf("%%1.1f = %1.1f\n",sample1);
    printf("%%9.1f = %9.1f\n",sample2);
    printf("%%9.2f = %9.2f\n",sample2);
    printf("%%9.3f = %9.3f\n",sample2);
    printf("%%9.4f = %9.4f\n",sample2);
    printf("%%9.5f = %9.5f\n",sample2);
    printf("%%9.6f = %9.6f\n",sample2);
    printf("%%9.7f = %9.7f\n",sample2);
    printf("%%9.8f = %9.8f\n",sample2);
    return(0);
}
```

Exercise 13-11: Type the source code from Listing 13-5 into your editor. It looks like a lot of work, but you can create the code quickly by using a lot of copy-and-paste.

The output from Exercise 13-11 helps illustrate the width and precision portions of the `%f` conversion character:

```
%9.1f =      34.5
%8.1f =     34.5
%7.1f =    34.5
```

```
%6.1f =   34.5
%5.1f =  34.5
%4.1f = 34.5
%3.1f = 34.5
%2.1f = 34.5
%1.1f = 34.5
%9.1f =      12.3
%9.2f =     12.35
%9.3f =    12.346
%9.4f =   12.3457
%9.5f =  12.34568
%9.6f = 12.345679
%9.7f = 12.3456793
%9.8f = 12.34567928
```

From this output, you can see how the width value "pads" the numbers on the left. As the width value decreases, so does the padding. However, when the width specified is wider than the original value, nonsense is displayed, as shown by the last two lines of output. This is because the width is beyond the limit of single precision.

Setting the output width

The *w* output width option is available to all the conversion characters, not just `%f`. The *width* is the minimum amount of space provided for output. When the output is less than the width, it's right-justified. When the output is greater than the width, the width value is ignored. Listing 13-6 provides examples using the `%s` placeholder.

LISTING 13-6: Messing with the Width

```
#include <stdio.h>

int main()
{
    printf("%%15s = %15s\n","hello");
    printf("%%14s = %14s\n","hello");
    printf("%%13s = %13s\n","hello");
    printf("%%12s = %12s\n","hello");
    printf("%%11s = %11s\n","hello");
    printf("%%10s = %10s\n","hello");
    printf(" %%9s = %9s\n","hello");
    printf(" %%8s = %8s\n","hello");
```

```
    printf(" %%7s = %7s\n","hello");
    printf(" %%6s = %6s\n","hello");
    printf(" %%5s = %5s\n","hello");
    printf(" %%4s = %4s\n","hello");
    return(0);
}
```

Exercise 13-12: Type the source code from Listing 13-6 into a new project. Build and run to examine the output, which looks like this:

```
%15s =           hello
%14s =          hello
%13s =         hello
%12s =        hello
%11s =       hello
%10s =      hello
 %9s =     hello
 %8s =    hello
 %7s =   hello
 %6s =  hello
 %5s = hello
 %4s = hello
```

As with the width option for floating-point numbers (refer to Listing 13-5), space is padded on the left when the width value is greater than the string. But when the width is less than the string's length, the full string is still output.

When the width value is specified for an integer, it can be used to right-align the output. For example:

```
printf("%4d",value);
```

This statement ensures that the output for `value` is right-justified and at least four characters wide. If `value` is fewer than four characters wide, it's padded with spaces on the left. That is, unless you stick a 0 in there:

```
printf("%04d",value);
```

In this case, the *printf()* function pads the width with zeros to keep everything four characters wide.

Exercise 13-13: Modify Exercise 13-1 so that the integer values' output is aligned. For example, the summary portion of the output should look something like this:

```
330 alphabetic characters
 70 blanks
  6 punctuation symbols
```

Aligning output

The width value in the conversion character aligns output to the right, known as *right justification.* But not everything is all right. Sometimes, you want left justification. To force the padding to the right side of the output, insert a minus sign before the width value in the `%s` conversion character. For example:

```
printf("%-15s",string);
```

This statement outputs the text in the array `string` justified to the left. If `string` is shorter than 15 characters, spaces are added to the right.

The source code in Listing 13-7 displays two strings. The first one is left-justified within a range of varying widths. The width gets smaller with each progressive *printf()* statement.

LISTING 13-7: Meeting in the Middle

```
#include <stdio.h>

int main()
{
    printf("%-9s me\n","meet");
    printf("%-8s me\n","meet");
    printf("%-7s me\n","meet");
    printf("%-6s me\n","meet");
    printf("%-5s me\n","meet");
    printf("%-4s me\n","meet");
    return(0);
}
```

Exercise 13-14: Copy the code from Listing 13-7 into your editor. Create the program and run it to see the alignment output demonstrated.

Exercise 13-15: Write a program that displays the first and last names of the first four presidents of the United States. You can express the names as literal values in the *printf()* statements. The names need to line up so that the output looks like this:

```
George Washington
John   Adams
Thomas Jefferson
James  Madison
```

Gently Down the Stream

The basic input/output functions in C are not interactive. They don't sit and wait for you to type text at the keyboard, which is the way you expect to use a computer program. But standard input in C isn't character based, it's stream based.

With *stream based* input, a program looks at the input as though it were poured out of a jug. All the characters, including Enter, march in, one after another. Only after a given chunk of text is received, or input stops altogether, does the stream end. This concept can be frustrating to any beginning C programmer.

Demonstrating stream input

Consider the code illustrated in Listing 13-8. It appears that the code reads input until the period is encountered. At this point, you would assume that input would stop, but doing so is not anticipating stream input.

LISTING 13-8: Foiled by Stream input

```
#include <stdio.h>

int main()
{
    int i;

    do
    {
        i = getchar();
        putchar(i);
    } while(i != '.');

    putchar('\n');
    return(0);
}
```

Exercise 13-16: Type the source code from Listing 13-8 into an editor. Build and run to try out the program. Type a lot of text and a period to see what happens.

Here's how it ran on my computer, with my typing shown in bold:

```
This is a test. It's only a test.
This is a test.
```

The program doesn't halt input after you type a period. The first line in the preceding example is the stream, like a fire hose shooting characters into the program. The program behaves properly, processing the stream and halting its display after the period is encountered. The Enter key serves as a break in the stream, which the program uses to digest input until that point.

Dealing with stream input

Despite the C language's stream orientation, ways do exist to create more-or-less interactive programs. You merely have to embrace stream input and deal with it accordingly.

The source code in Listing 13-9 should be pretty straightforward to you. The *getchar()* function fetches two characters and then the characters are displayed on Line 11.

LISTING 13-9: Fishing for Characters in the Stream

```
#include <stdio.h>

int main()
{
    int first,second;

    printf("Type your first initial: ");
    first = getchar();
    printf("Type your second initial: ");
    second = getchar();
    printf("Your initials are '%c' and '%c'\n", first,second);
    return(0);
}
```

Exercise 13-17: Type the source code from Listing 13-9 into your editor. Line 11 in the listing is split so that it doesn't wrap; you don't have to split the line in your editor. Build and run using your initials as input.

Here's the output I saw, with my typing shown in bold:

```
Type your first initial: D
Type your second initial: Your initials are 'D' and '
'
```

Like you, I never got a chance to type my second initial. The stream included the Enter key press, which the program accepted as input for the second *getchar()* function. That character, \n, is displayed in the output between the single quotes.

How do you run the program? Simple: Type both initials at the first prompt:

```
Type your first initial: DG
Type your second initial: Your initials are 'D' and 'G'
```

Of course, this isn't what the code asks for. So how do you fix it? Can you think of a solution using your current programmers' bag o' tricks?

Don't give up!

The solution I would use is to devise a function that returns the first character in the stream and then swallows the rest of the characters until the \n is encountered. This function appears in Listing 13-10:

LISTING 13-10: A Single-Character Input Function, *getch()*

```
int getch(void)
{
    int ch;

    ch = getchar();
    while(getchar()!='\n')
        ;
    return(ch);
}
```

To wrap your brain around stream input, consider that the *while* loop in Listing 13-10 spins through all text in the stream until a newline is encountered. Then the first character in the stream, grabbed at Line 5, is returned from the function.

Exercise 13-18: Modify the source code from Exercise 13-17 so that the *getch()* function illustrated in Listing 13-10 is used to gather input. Build and run to ensure that the output is what the user anticipates.

TIP

If you want truly interactive programs, I recommend that you look into the NCurses library, which extends C's capability to output and input text. NCurses lets you create full-screen text programs that are immediately interactive.

» Declaring structure variables

» Assigning values to a structure variable

» Building structure arrays

» Putting one structure inside another

Chapter 14
Structures, the Multivariable

Individual variables are perfect for storing single values. When you need more of one type of a variable, you declare an array. For data that consists of several different types of variables, you mold the variable types into something called a *structure*. It's the C language's method of creating a variable buffet.

Hello, Structure

I prefer to think of the C language structure as a *multivariable*, or several variables rolled into one. You use structures to store or access complex information. That way, you can keep various *int*, *char*, *float* variables, and even arrays, all in one neat package.

Introducing the multivariable

Some things just belong together — like your name and address or your bank account number and all the money that's supposedly there. You can craft such a relationship in C by using parallel arrays or specifically named variables. But this method is clunky. A better solution is to employ a structure, as demonstrated in Listing 14-1.

LISTING 14-1: One Variable, Many Parts

```
#include <stdio.h>

int main()
{
    struct player
    {
        char name[32];
        int highscore;
    };
    struct player xbox;

    printf("Enter the player's name: ");
    scanf("%s",xbox.name);
    printf("Enter their high score: ");
    scanf("%d",&xbox.highscore);

    printf("Player %s has a high score of %d\n",
            xbox.name,xbox.highscore);
    return(0);
}
```

Exercise 14-1: Without even knowing what the heck is going on, type Listing 14-1 into your editor to create a new program. Build and run.

Here's how the code in Listing 14-1 works:

Lines 5 through 9 declare the `player` structure. This structure has two members — a *char* array (string) and *int* — declared just like any other variables, in Lines 7 and 8.

Line 10 declares a new variable for the `player` structure, `xbox`.

Line 13 uses *scanf()* to fill the `name` member for the `xbox` structure variable with a string value.

Line 15 uses *scanf()* to assign a value to the `highscore` member in the `xbox` structure.

The structure's member values are displayed at Line 17 by using a *printf()* function. The statement is split between Lines 17 and 18, with the arguments for *printf()* on Line 18.

Understanding *struct*

A structure is a collection. Think of it as a frame that holds multiple variable types. In many ways, a structure is similar to a record in a database. For example:

```
Name
Age
Gambling debt
```

These three items can be fields in a database record, but they can also be members in a structure: `Name` would be a string; `Age`, an integer; and `Gambling Debt`, an unsigned floating-point value. Here's how such a record would look as a structure in C:

```
struct record
{
    char name[32];
    int age;
    float debt;
};
```

struct is a C language keyword that introduces a structure. It's a definition that says, "This structure holds the following data types."

`record` is the name of the new structure being created. It's not a variable — it's a structure type.

Within the curly brackets dwell the structure's members, the variables contained in the named structure. The `record` structure type contains three member variables: a string `name`, an *int* named `age`, and a *float* value, `debt`.

The *struct* keyword only defines a structure type, setting its contents. To use the structure, you must declare a structure variable of the structure type you created. For instance:

```
struct record human;
```

This statement declares a new variable `human` of the `record` structure type.

Structure variables can also be declared when you define the structure itself. For example:

```
struct record
{
    char name[32];
    int age;
    float debt;
} human;
```

These statements define the `record` structure *and* declare a `record` structure variable, `human`. Multiple variables of that structure type can also be created:

```
struct record
{
    char name[32];
    int age;
    float debt;
} bill, mary, dan, susie;
```

Four `record` structure variables are created in this example.

To access members in a structure variable, you use a period, which is the *member operator*. It connects the structure variable name with a member name. For example:

```
printf("Victim: %s\n",bill.name);
```

This statement references the `name` member in the `bill` structure variable. A *char* array, it's used in your code like any other *char* array. Other members in the structure variable are used like their individual counterparts as well:

```
dan.age = 32;
```

In this example, the `age` member in the structure variable `dan` is set to the value `32`.

Exercise 14-2: Modify the source code from Listing 14-1 so that another member is added to the player structure, a *float* value indicating `hours` played. Spruce up the rest of the code so that the new value is input and displayed.

Filling a structure

As with other variables, you can initialize a structure variable when it's created. You first define the structure type and then declare a structure variable with its member values preset. Ensure that the preset values match the order and type of members defined in the structure, as shown in Listing 14-2.

LISTING 14-2: Declaring an Initialized Structure

```
#include <stdio.h>

int main()
{
    struct president
    {
        char name[40];
        int year;
    };
    struct president first = {
        "George Washington",
        1789
    };

    printf("The first president was %s\n",first.name);
    printf("He was inaugurated in %d\n",first.year);

    return(0);
}
```

Exercise 14-3: Create a new program by typing the source code from Listing 14-2 into the editor. Build and run.

You can also declare a structure and initialize it in one statement:

```
struct president
{
    char name[40];
    int year;
} first = {
    "George Washington",
    1789
};
```

Exercise 14-4: Modify your source code from Exercise 14-3 so that the structure and variable are declared and initialized as one statement.

Exercise 14-5: Add another `president` structure variable, `second`, to your code, initializing that structure with information about the second president, John Adams, who was inaugurated in 1797. Display the contents of both structures.

Making an array of structures

Creating individual structure variables, one after the other, is as boring and wasteful as creating a series of any individual variable type. The solution for multiple structures is the same as for multiple individual variables: an array.

A structure array is declared like this:

```
struct scores player[4];
```

This statement declares an array of `scores` structures. The array is named `player`, and it contains four structure variables as its elements.

The structures in the array are accessed by using a combination of array and structure notation. For example:

```
player[2].name
```

The variable in the preceding line accesses the `name` member in the third element (2) in the `player` structure array. Yes, that's the third element because the first element would be referenced like this:

```
player[0].name
```

REMEMBER

Arrays start numbering with the element `0`, not element `1`.

Line 10 in Listing 14-3 declares an array of four `scores` structures. The array is named `player`. Lines 13 through 19 fill each structure in the array. Lines 21 through 27 display each structure's member values.

LISTING 14-3: Arrays of Structures

```
#include <stdio.h>

int main()
```

```
{
    struct scores
    {
        char name[32];
        int score;
    };
    struct scores player[4];
    int x;

    for(x=0;x<4;x++)
    {
        printf("Enter player %d: ",x+1);
        scanf("%s",player[x].name);
        printf("Enter their score: ");
        scanf("%d",&player[x].score);
    }

    puts("Player Info");
    printf("#\tName\tScore\n");
    for(x=0;x<4;x++)
    {
        printf("%d\t%s\t%5d\n",
            x+1,player[x].name,player[x].score);
    }
    return(0);
}
```

Exercise 14-6: Type the source code from Listing 14-3 into your editor. Build and run the program.

Exercise 14-7: Add code to Listing 14-3 so that the display of structures is sorted with the highest score listed first. Yes, you can do this. Sorting an array of structures works just like sorting any other array. Review Chapter 12 if you suddenly lose your nerve.

TIP

Here's a hint, just because I'm a nice guy. Line 28 of my solution looks like this:

```
player[a]=player[b];
```

You can swap structure array elements just as you can swap any array elements. You don't need to swap the structure variable's members.

Weird Structure Concepts

I'll admit that structures are perhaps the weirdest type of variable in the C language. The two steps required to create them are unusual, but the dot method of referencing a structure's member always seems to throw off beginning programmers. If you think that, beyond those two issues, structures couldn't get any odder, you're sorely mistaken.

Putting structures within structures

It's true that a structure holds C language variables. It's also true that a structure is a C language variable. Therefore, it follows that a structure can hold another structure as a member. Don't let this type of odd thinking confuse you. Instead, witness the example shown in Listing 14-4.

LISTING 14-4: A Nested Structure

```
#include <stdio.h>
#include <string.h>

int main()
{
    struct date
    {
        int month;
        int day;
        int year;
    };
    struct human
    {
        char name[45];
        struct date birthday;
    };
    struct human president;

    strcpy(president.name,"George Washington");
    president.birthday.month = 2;
    president.birthday.day = 22;
    president.birthday.year = 1732;
```

```
    printf("%s was born on %d/%d/%d\n",
            president.name,
            president.birthday.month,
            president.birthday.day,
            president.birthday.year);

    return(0);
}
```

Listing 14-4 declares two structure types: `date` at Line 6 and `human` at Line 12. Within the `human` structure's declaration, at Line 15 you see the `date` structure variable `birthday` declared. That's effectively how one structure is born inside another.

Line 17 creates a `human` structure variable, `president`. The rest of the code fills this structure's members with data. The method for accessing a nested structure's members is shown in Lines 20 through 22.

Exercise 14-8: Type the source code from Listing 14-4 into your editor. Build and run the program.

Exercise 14-9: From your solution for Exercise 14-8, replace the `name` member in the `human` structure with a nested structure. Name that structure `id` and have it contain two *char* array members for storing an individual's first and last names: `first` and `last`. If you do everything correctly, the reference to the president's name will be the variables `president.name.first` and `president.name.last`. Be sure to assign values to these variables in your code and display the results.

Passing a structure to a function

As a type of variable, it's entirely possible for a function to accept a structure as an argument and return it as a value. However, this process requires that the structure be declared externally, as a global variable. The reason is that the structure's definition must be available to all functions in the source code.

The topic of global variables is avoided until Chapter 16. It's not that complex, but learning about passing and returning structures can be delayed until then.

» Working with the command line

» Specifying *main()*'s arguments

» Understanding the *exit()* function

» Running programs with *system()*

Chapter 15
Life at the Command Prompt

Before computers went graphical, the text screen was as high-tech as computers went. Visually, everything was plain and dull, and the most exciting computer games involved a lot of reading. If Facebook were invented back then, it would be all book and no face.

Way back when, computer life centered around the command prompt. Directions were typed, and output was plain text. That's the environment in which the C language was born, and to some extent, where it still exists today.

Conjure a Terminal Window

Whether you're using Windows, Mac OS X, Linux, or a Unix variant, you can still bring forth a terminal window, in which you can witness the breathtaking starkness of the command prompt. It's a text-only environment, and, in fact, it's the environment in which each and every program you've coded in this book is perfectly at home.

Starting a terminal window

Figure 15-1 illustrates the Ubuntu Linux bash shell in Windows 10. Similar command prompt or terminal windows are available for other platforms: On the Macintosh, start the Terminal app. For Linux, start the terminal window.

```
Ubuntu
Hello dang
~$ cd Dan/prog/c
c$ ls
 02_01-ide_debug   0425.c         debug2      getpass.o                 test.c
 02_02             0425.exe       debug3     'inverse text test.c'      test.exe
 02_03             0425.o         ex0101     'inverse text test.o'      test.o
 02_05             Test           ex1010      inverse.exe
 0418             'debug test'    getpass.c   kill
c$
```

FIGURE 15-1: Text mode in a terminal window.

Open the Terminal app on your computer to witness its nongraphical glory.

- In Windows, choose Ubuntu from the Start button menu.
- On the Mac, in the Finder, press Command+Shift+U to view the Utilities folder. Open the Terminal app.
- In a Linux or Unix GUI environment, locate and launch the Terminal, Term, or Xterm program.

The command prompt is both cryptic and powerful. In fact, programmers at Microsoft use a command-prompt environment called PowerShell to configure Windows. For the configurations just mentioned in the list, you can compile and run all the programs presented in this book. In fact, you can use the terminal window to do everything: edit, build, and run — just like back in the 1980s!

To close the command prompt or terminal window, type the **exit** command and press the Enter key.

TIP

- Code::Blocks for Windows uses the old DOS command prompt, the *cmd* program, to output its text mode programs. This environment is rather limited, which is why I recommend obtaining the Ubuntu shell for Windows 10.
- Using the Ubuntu shell in Windows 10 involves more than just obtaining the app from the Microsoft Store. You must also configure Windows per the directions offered. Don't worry! If you screw up, the error message explains what to do.

Running code in text mode

The programs you create in this book all run in Text mode. If you're using Code::Blocks, you see an output window, but you can also run your programs directly at the DOS command prompt. The trick is finding the program, which requires a little command-prompt acumen.

Why do this?

Because it's the best way to test programs that access the *main()* function's arguments, which are obtained from the command prompt. Heed these directions:

1. **Open the terminal window.**

   Refer to the directions in the preceding section.

   The command prompt or terminal window opens to your home folder, which is called a *directory* in text mode. The home directory most likely doesn't contain your C programs, so you must change to another directory.

2. **Use the *cd*, change *d*irectory, command to switch to the folder you created for storing this book's projects.**

   For example, I save my C programming projects in the `prog/c` directory. For this book, source code files dwell in the `cprog` subdirectory. The shell command I type to visit this directory is

   ```
   cd ~/prog/c/cprog
   ```

   Type a space after the *cd* command and then type the pathname, such as my `~/prog/c/cprog` (where ~ is the shortcut to your account's home directory).

Type the pathname on your own system to visit the proper directory. If you're using Code::Blocks, the project folder is named after the project you've saved, such as `ex1409` for the last project in Chapter 14. To run the program that's created, you must change to the project's `bin` folder:

```
cd ex1409/bin/release
```

The current directory now contains the executable program file for your project.

3. **To run the program, type its name at the command prompt.**

For example, the default output name for the *cc*, *gcc*, and *clang* compiler is `a.out`. Type this command:

```
./a.out
```

The `./` directs the shell to look for an executable file in the current directory, `a.out`. Otherwise, you follow the `./` with the program file's name.

Follow these steps to run any program you've created at the command prompt. Alternative ways exist to specify the arguments in an IDE. These methods are covered elsewhere in this chapter.

TIP

FINDING WINDOWS FILES IN UBUNTU

If you're using Ubuntu in Windows 10, be aware that it uses a separate file system from Windows. To access files in the Windows file system, at the bash shell you must use this command:

```
cd /mnt/c
```

The `/mnt/c` folder is a mount point for drive C, which is where Windows is normally installed. From there, you can change to the `Users` directory, and then to your user profile folder. Then you can change to the directory (folder) where you keep program files.

This technique offers a good solution for using the Linux bash shell for coding, but be aware that you shouldn't venture in the other direction: Don't use Windows to find the Ubuntu file system and use the File Explorer to manage files there. If you do, the files may not synchronize properly, which leads to more problems you don't want.

Arguments for the *main()* Function

Back in the old days, programs featured command-line options or switches. For example, to compile and link a C program, you type something like this:

```
cc ex1501.c -o ex1501
```

The three tidbits of text after the *cc* command are options or switches. With the *cc* program, these text tidbits are accessed as arguments to the *main()* function. A program can read these arguments to perform certain actions. Even today, when the world runs graphical operating systems, command-line arguments are relevant. All you need to do in your code is examine the arguments to the *main()* function.

Reading the command line

Pretend that it's 1987 and you're writing a program that says "Hello" to the user by name. The way you get the user's name is to have your code swallow the first chunk of text that appears after the program name at the command line. That code may look something like Listing 15-1.

LISTING 15-1: Well, Hello There!

```
#include <stdio.h>

int main(int argc, char *argv[])
{
    if(argc>1)
        printf("Greetings, %s!\n",argv[1]);
    return(0);
}
```

Line 3 in Listing 15-1 is different from the ones you see earlier in this book. Instead of being empty, the *main()* function now shows its two arguments — `argc` and `*argv[]` — in its parentheses.

Line 5 uses the *int* value `argc` to determine whether any additional items were typed after the program name at the command prompt.

Line 6 uses the string value (*char* array) `argv[1]` to display the first item after the program name at the command prompt.

Exercise 15-1: Type the source code from Listing 15-1 into a new project. Build and run.

The program displays no output unless a command-line argument is specified. To do so at the command prompt:

```
./a.out Jonah
```

The name of the program is `a.out`. The `./` prefix directs the command interpreter to look for the file in the current directory. This command is followed by a space and then the first command-line argument, `Jonah`. The program outputs this message:

```
Greetings, Jonah!
```

If you're using Code:: Blocks, follow these steps to set a command-line argument:

1. **Choose Project ⇨ Set Programs' Arguments.**

   This command is available when you start a Code::Blocks project. Upon success, you see the Select Target dialog box, as shown in Figure 15-2.

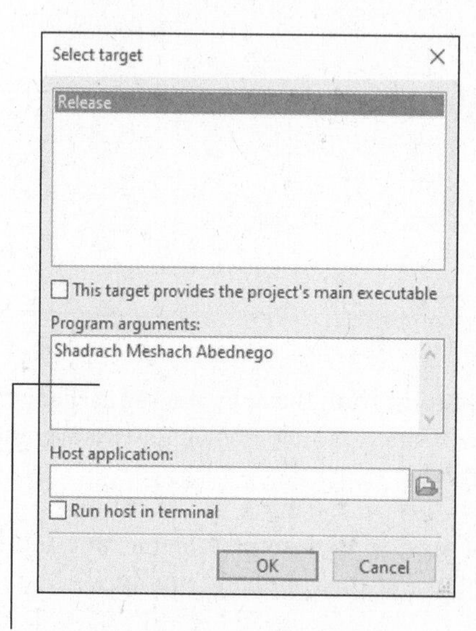

FIGURE 15-2: Setting command-line arguments in Code::Blocks.

2. **Type command-line text in the Program Arguments portion of the Select Target dialog box.**

   Use Figure 15-2 as your guide. Type the arguments just as though they were typed at the command prompt — spaces, dashes, jots, and tittles.

3. **Click the OK button.**
4. **Run your program again to see its output given the command-line arguments.**

If you're coding at the command prompt in DOS, you can run the program like this at the DOS prompt:

```
ex1501 Shadrach
```

Press the Enter key to run.

The code from Listing 15-1 uses only the first command-line argument, so if you type more items, they're ignored. For example:

```
ex1501 Shadrach Meshach Abednego
```

In the preceding line, only Shadrach's name appears in the output.

Understanding *main()*'s arguments

When you don't plan on your program accepting any command-line arguments, you can leave the *main()* function's parentheses empty. Like this:

```
int main()
```

When arguments are used in your code, they must be declared. Using them looks like this:

```
int main(int argc, char *argv[])
```

`argc` is the argument count value. It's an integer that ranges from 1 through however many items were typed after the program name at the command prompt.

`*argv[]` is an array of *char* pointers. You can think of it instead as an array of strings, which is how it can be used in your code.

The code in Listing 15-2 merely counts the number of arguments typed at the command line. That value, `argc`, is output.

LISTING 15-2: Argument Counter

```
#include <stdio.h>

int main(int argc, char *argv[])
{
    printf("You typed %d arguments.\n",argc);
    return(0);
}
```

Exercise 15-2: Type the preceding source code. Build and run by typing no arguments.

The *main()* function receives information about the command-line argument directly from the operating system. The command line is evaluated, and arguments are tallied and referenced. The tally appears as `argc`, and the references are stored in the `argv[]` array.

When no arguments are typed — in Code::Blocks, that means the Program Arguments window remains empty (refer to Figure 15-2) — you see this output:

```
You typed 1 arguments.
```

That's because the program name itself is the first argument. You can prove it by adding a single line to the code:

```
printf("Argument one is %s.\n",argv[0]);
```

Exercise 15-3: Modify your source code by adding the preceding line, inserting it after the first *printf()* statement. Build and run.

The program's output now displays the program's name — most likely, a full path to the program, which is accurate but a bit of overkill.

Exercise 15-4: Modify the code again, this time adding a *for* loop to work through all the arguments and displaying each one. For example, the output may look like this:

```
begc4d$ ./ex1504 Shadrach Meshach Abednego
Arg#1 = ./ex1504
Arg#2 = Shadrach
Arg#3 = Meshach
Arg#4 = Abednego
```

REMEMBER

The first argument is element zero of the `argv[]` array.

Time to Bail

Information can get into your program via command-line arguments. Information gets back out thanks to the *return* statement. That's the primary, but not the only, way a program bails out when it's done.

Quitting the program

Your program quits when the *main()* function encounters the *return* statement. Traditionally, that statement appears at the end of the function, but it doesn't always need to go there. Further, you can use the *exit()* function to leave the program at any time, even within a function other than *main()*.

The *exit()* function gracefully quits a program, tying up any loose ends, tucking variables into bed, and so on. In Listing 15-3, this function is used at Line 17 to leave the program in the *sub()* function.

LISTING 15-3: There Must Be Some Way Out of Here

```
#include <stdio.h>
#include <stdlib.h>

void sub(void);

int main()
{
    puts("This program quits before it's done.");
    sub();
    puts("Or was that on purpose?");
    return(0);
}
```

(continued)

LISTING 15-3: (continued)

```
void sub(void)
{
    puts("Which is the plan.");
    exit(0);
}
```

You must include the `stdlib.h` header file to use the *exit()* function, and it consumes an *int* value as an argument for the exit status, similar to the value passed by *return* in the *main()* function.

Exercise 15-5: Type the source code from Listing 15-3 into your edit. Build and run the program.

Running another program

The *system()* function directs your program to run another program or to issue a command. For example:

```
system("blorf");
```

The preceding statement directs the operating system to issue the `blorf` command, running whatever program has that name or carrying out whatever actions the `blorf` command dictates.

After running the command, control returns to your program, which continues with the statement following the *system()* function.

Listing 15-4 contains two *system()* functions; your code needs only one. Use the first *system()* statement if you're using Windows; use the second statement if you're using anything else. Or you can just comment out the statement rather than delete it.

LISTING 15-4: Clearing Things Up

```
#include <stdio.h>
#include <stdlib.h>

int main()
{
    printf("Press Enter to clear the screen:");
    getchar();
```

```
    system("cls");        /* Windows only */
    system("clear");      /* Mac - Unix */
    puts("That's better");
    return(0);
}
```

Line 2 includes the `stdlib.h` header file, which is required for the *system()* function to work. Ensure that the command to be run is a string literal enclosed in double quotes or is represented by a *char* array (string).

Exercise 15-6: Create a new project by using the source code shown in Listing 15-4. Build and run.

- Using *typedef* to create new variables
- Making static variables
- Creating a global variable
- Working with structures and functions
- Exploring enumerated constants

Chapter 16
Variable Nonsense

You have more to learn about C language variables than knowing the keywords *int*, *char*, *float*, and *double*. Yes, I'm including *signed*, *unsigned*, *long*, and anything else you may already know in that list. The reason is that the variable is a big part of C. Choosing the right data type and using it properly can make or break a program.

Variable Control

That which is called a variable by any other name would still be a variable. That is, unless you mess with the variable's data type in your code by changing it into another type, giving it a new name altogether, or casting a spell upon the variable to meet your needs, benevolent or not.

Typecasting into disbelief

When is a *float* variable not a *float*? When it's typecast into an *int*, of course. This trick is made possible in C by using the typecast. For example:

```
(int)debt
```

In the preceding line, the *float* variable `debt` is typecast to an *int* value. The *int* in parentheses directs the compiler to treat the value of `debt` as an integer.

Why would anyone want to do that?

Sometimes a function requires a specific data type and the type isn't available. Rather than juggle several data types in one program, you just typecast a variable into the data type you desire. This trick is often necessary, as shown in Listing 16-1.

LISTING 16-1: That's Not Right

```
#include <stdio.h>

int main()
{
    int a,b;
    float c;

    printf("Input the first value: ");
    scanf("%d",&a);
    printf("Input the second value: ");
    scanf("%d",&b);
    c = a/b;
    printf("%d/%d = %.2f\n",a,b,c);
    return(0);
}
```

Exercise 16-1: Type the source code from Listing 16-1 into your editor. Build and run.

Here's a sample run with my input in bold:

```
Input the first value: 3
Input the second value: 2
3/2 = 1.00
```

Obviously, I'm incorrect in my assumption that 3 ÷ 2 would somehow work out to 1.50. If the computer says it's 1.00, the computer must be correct.

Or perhaps the computer is merely confused because, in Line 12 of the source code, two *int* values are divided and the result is assigned to a *float*. This operation doesn't quite work, however, because dividing an integer by an integer in C yields an integer as the result. The value 1 is the closest integer value to 1.50. So even though the computer is wrong, it's doing exactly what it was told to do.

Exercise 16-2: Modify your source code, changing Line 12 to read

```
c = (float)a/(float)b;
```

Save the change. Build and run using the same values as just shown. Here's the new output:

```
Input the first value: 3
Input the second value: 2
3/2 = 1.50
```

Better. That's because you *typecast* variables a and b in the equation, temporarily allowing the compiler to treat them as floating-point numbers. Therefore, the result is what it should be.

- To typecast a variable, prefix it with the data type desired, enclosed in parentheses:

  ```
  (float)boat;
  ```

  In this line, integer variable boat is typecast to a *float* value.

- A typecast is different from a *typedef*, covered in the next section.

Creating new things with *typedef*

You can get into loads of trouble with the *typedef* keyword. Beware! It's powerful. Heck, it's more than a keyword — it's a spell! It can cast normal C words and operators from their consistent selves into all sorts of mischief.

And because my editor needs a reference in the text to this next code sample, see Listing 16-2.

LISTING 16-2: The Perils of *typedef*

```
#include <stdio.h>

typedef int stinky;

stinky main()
```

(continued)

LISTING 16-2: (continued)

```
{
    stinky a = 2;

    printf("Everyone knows that ");
    printf("%d + %d = %d\n",a,a,a+a);
    return(0);
}
```

In Listing 16-2, the *typedef* statement at Line 3 creates a new data type, named `stinky`. It's created as a synonym for the keyword *int*. Anywhere you can use *int* in the code, you can use the word `stinky` instead, as shown on Lines 5 and 7.

Exercise 16-3: Use the source code from Listing 16-2 to create a new program, demonstrating that a `stinky` variable type is the same as an *int*. Build and run.

Granted, the example in Listing 16-2 is rather silly; no serious programmer would set up a real program like that. Where *typedef* is used most often is in defining structures. The *typedef* statement helps to reduce the chunkiness of this activity.

For example, Exercise 14-9 (from Chapter 14) directs you to declare two structures nested in a third. Listing 16-3 shows how that operation works, given a knowledge of structures (from Chapter 14):

LISTING 16-3: Creating a Structure the Traditional Way

```
struct id
{
    char first[20];
    char last[20];
};

struct date
{
    int month;
    int day;
    int year;
};

struct human
{
    struct id name;
    struct date birthday;
};
```

Listing 16-4 shows how the declarations take place if you were to use *typedef* to simplify the structures:

LISTING 16-4: Using *typedef* to Define a Structure

```
typedef struct id
{
    char first[20];
    char last[20];
} personal;

typedef struct date
{
    int month;
    int day;
    int year;
} calendar;

struct human
{
    personal name;
    calendar birthday;
};
```

In this listing, the structure `id` is *typedef*'d to the name `personal`. This isn't a variable name; it's a *typedef*. It's the same as saying, "All references to `struct id` are now the same as the name `personal`."

Likewise, the structure `date` is *typedef*'d to `calendar`. Finally, in the declaration of the structure `human`, the *typedef* names are used instead of the more complex structure definitions.

Exercise 16-4: Modify the source code from the project you create in Exercise 14-9 (in Chapter 14) to use *typedef*, as shown in Listing 16-4. Build and run.

It can be argued that using *typedef* doesn't make your code any more clear than had you just used good variable names and well-formatted text. For example, I don't use *typedef* because I don't want to have to remember what I've defined. But you will encounter other code that uses *typedef*. Don't let it freak you out.

» To cast the typedef spell, use this format:

```
typedef data_type new_type
```

Follow the *typedef* keyword with an existing C data type, *data_type*, and then the synonym for the data type, *new_type*. Use the same rules for naming a variable as you would for the *new_type* definition.

- One advantage of using *typedef* with a structure is that it saves you from typing the word *struct* too many times.
- Just about any data type you encounter in the C library other than a keyword, such as *time_t*, *size_t*, or *st_mode*, is most likely a *typedef*, declared in a header file. The library documentation (*man* page) explains the underlying data type, but you should use the *typedef*-defined name in your code to declare any required variables. For example:

```
time_t now;
```

  The preceding statement creates a variable now of the *time_t* data type, defined in the time.h header file. See Chapter 21 for more information on C library time functions.

- The rare times the underlying data type for a *typedef* must be known is when using a *typedef* variable in a *printf()* statement. The conversion character must match the actual data type. Fortunately, modern compilers, like *clang*, report the correct conversion character to use.

WARNING

- Where programmers get into trouble with *typedef* and structures is when creating a linked list. I repeat this warning in Chapter 20, which covers linked lists.

Making *static* variables

Variables used within a function are *local* to that function: Their values are used and then discarded when the function is done. Listing 16-5 demonstrates the concept.

LISTING 16-5: Don't Give Me No *static*

```
#include <stdio.h>

void proc(void);

int main()
{
    puts("First call");
    proc();
    puts("Second call");
```

```
    proc();
    return(0);
}

void proc(void)
{
    int a;

    printf("The value of variable a is %d\n",a);
    printf("Enter a new value: ");
    scanf("%d",&a);
}
```

In Listing 16-5, variable `a` in the *proc()* function doesn't retain its value. The variable is initialized only by the *scanf()* function at Line 20. Otherwise, the variable contains junk information.

Exercise 16-5: Type the source code from Listing 16-5 into your editor. Build the code. Ignore the "uninitialized" warning that may appear. Run the program.

On my computer, the output looks like this:

```
First call
The value of variable a is 0
Enter a new value: 6
Second call
The value of variable a is 0
Enter a new value: 6
```

Despite my attempts to assign 6 to variable `a`, the program always forgets. So much for that. Or is it?

Exercise 16-6: Modify the source code from Listing 16-5, editing Line 16 to read:

```
static int a;
```

Build and run to test the output. Here's what I see:

```
First call
The value of variable a is 0
Enter a new value: 6
Second call
The value of variable a is 6
Enter a new value: 5
```

VARIABLE KEYWORD ROUNDUP

The C language features a few keywords that puzzle beginners and advanced users alike. Here's the roundup of various variable classifier, quantifier, and what-have-you-ifier keywords related to variables.

auto: The *auto* keyword is a storage class specifier. It's a holdover from the ancient B programming language, which defines all variables used in a function that aren't constants or *static*. Because this definition means pretty much every variable declaration, the *auto* variable classification is assumed by the compiler and, therefore, this keyword prefix isn't required.

const: The *const* keyword is a quantifier used to create a constant or unchanging value. Refer to Chapter 6.

enum: The *enum* keyword creates an *enumerated* list of constants. See the section "Enumerating," later in this chapter.

extern: Like *auto*, the *extern* keyword is a storage class specifier, referencing variables declared outside of a function. It's covered in Chapter 24, though the concept of external variables is covered elsewhere in this chapter.

register: Another storage class specifier, a variable declared as a *register* (followed by the data type and variable name) is to be stored directly in a CPU register. Modern compilers are highly optimized, which makes this keyword unnecessary.

static: The *static* keyword declares a storage class that isn't discarded when a function terminates. It's covered elsewhere in this chapter.

union: A *union* is a complex data construction, similar to a structure. Unlike a structure, the union can hold a variety of data types in the same memory space, though the code can access only one type at a time. Unions are considered a security risk because they define a storage area with an inconsistent data type.

volatile: This scary keyword is a quantifier, often called the opposite of the *const* keyword. Like *register*, *volatile* was once used to optimize code by telling the compiler which variables would change most often. Such optimization is no longer necessary with modern C compilers.

Because the variable was declared as *static*, its value is retained between function calls.

- As a bonus, *static* variable declarations are initialized to zero.
- The keyword *static* defines a storage class. See the nearby sidebar, "Variable keyword roundup."
- You need not declare variables as *static* unless their values must be retained each time the function is called. This situation crops up from time to time. Also refer to the later section "Using external variables."
- Values returned from a function need not be declared *static*. When you return a value, such as

  ```
  return(a);
  ```

  only the variable's value is returned, not the variable itself. The only challenge with this approach occurs when allocating memory as the buffer. The buffer's data is discarded and should be declared *static* if you want the calling function to access it. See Chapter 19.

Variables, Variables Everywhere

Sometimes, a variable must be like cellular phone service: available everywhere and accessed from any function. Often referred to as a *global* variable, in C it's officially known as an external variable.

Using external variables

External variables solve specific problems by making the variable declaration available to all functions in the source code file. Any function can access the variable. It doesn't have to be passed to or returned from a function.

Listing 16-6 shows how external variables are declared and used. The variables `age` and `feet` are external. Both are used in, and manipulated by, all functions in the source code.

LISTING 16-6: Tossing Your Age Around

```
#include <stdio.h>

void half(void);
void twice(void);

int age;
float feet;

int main()
{
    printf("How old are you: ");
    scanf("%d",&age);
    printf("How tall are you (in feet): ");
    scanf("%f",&feet);
    printf("You are %d years old and %.1f feet tall.\n",
  age,feet);
    half();
    twice();
    printf("But you're not really %d years old or %.1f feet
  tall.\n",age,feet);
    return(0);
}

void half(void)
{
    float a,h;

    a=(float)age/2.0;
    printf("Half your age is %.1f.\n",a);
    h=feet/2.0;
    printf("Half your height is %.1f.\n",h);
}

void twice(void)
{
    age*=2;
    printf("Twice your age is %d.\n",age);
    feet*=2;
    printf("Twice your height is %.1f\n",feet);
}
```

Line 6 declares the external *int* variable `age`, and Line 7 creates external *float* variable `feet`. These variables are declared outside of any function, up there in `#include`, `#define`, and prototyping land. The variables are available to every function. Even when those values are changed in the *twice()* function, the *main()* function uses the new values.

Be aware that two *printf()* statements in the *main()* function wrap their text in Listing 16-6. You don't need to wrap those statements in a text editor; just type each one on a single line.

Exercise 16-7: Type the source code for Listing 16-6 into your editor, creating a new program. Build and run.

Though both `age` and `feet` variables are external, the *extern* keyword isn't used to declare them. This keyword is required only when linking multiple source code files (modules) into a single program. In this configuration, the keyword identifies external variables available in another source code file but not declared in the current file. Chapter 24 explains how to use the *extern* keyword in full detail.

WARNING

» Don't be lazy about using global variables! If you can pass a value to a function, do so! It's proper. Too many indolent programmers declare all their variables external to "solve the problem." This approach is sloppy and improper.

TIP

» Good examples of global variables include information that all functions in the program must access. For example, status settings, user information, and other items too difficult to pass to every function make excellent candidates for external variables.

» In my programming history, it's rare that code begs for an external variable. Only when I can't figure out any other way to make the information available do I declare a variable globally.

Creating an external structure variable

A situation where external variables are completely necessary occurs when passing a structure to a function. In this instance, you must declare the structure externally so that all functions can access variables of the given structure type. Further, the structure must be declared before any function prototypes for which the structure appears as an argument.

Don't let the massive length of Listing 16-7 intimidate you! Most of the "real" programs you eventually write will be far longer!

LISTING 16-7: Passing a Structure to a Function

```
#include <stdio.h>
#include <stdlib.h>
#include <time.h>

struct bot {
    int xpos;
    int ypos;
};

struct bot initialize(struct bot b);

int main()
{
    const int size = 5;
    struct bot robots[SIZE];
    int x;

    srand((unsigned)time(NULL));

    for(x=0;x<SIZE;x++)
    {
        robots[x] = initialize(robots[x]);
        printf("Robot %d: Coordinates: %d,%d\n",
                x+1,robots[x].xpos,robots[x].ypos);
    }
    return(0);
}

struct bot initialize(struct bot b)
{
    int x,y;

    x = rand();
    y = rand();
    x%=20;
    y%=20;
    b.xpos = x;
    b.ypos = y;
    return(b);
}
```

To pass a structure to a function, the structure must be declared externally, which happens between Lines 5 and 8 in Listing 16-7. This declaration is required before the function is prototyped, which takes place at Line 10.

The *initialize()* function runs from Lines 39 through 40. The structure is passed to the function and returned. Note that the structure variable must be fully defined as the argument. On Line 29, the structure is given the variable name `b` inside the function.

The *return* statement at Line 39 passes the structure back to the calling function. Indeed, the *initialize()* function is defined as a structure `bot` data type. This type is the value it returns.

Exercise 16-8: Screw your courage to the sticking place, and type all those lines of source code from Listing 16-7 into your editor. Build and run.

The output demonstrates how the structure array was passed (one element at a time) to a function, modified in the function, and then returned.

Enumerating

The *enum* keyword need not baffle you. Instead, bathe yourself in its warm comfort whenever you must declare a series of sequentially numbered constants. Listing 16-8 shows an example that might be familiar to you.

LISTING 16-8: Verifying That Input Value

```
#include <stdio.h>

int verify(int check);

int main()
{
    int s;

    printf("Enter a value (0-100): ");
    scanf("%d",&s);
    if(verify(s))
    {
        printf("%d is in range.\n",s);
    }
    else
```

(continued)

LISTING 16-8: *(continued)*

```
    {
        printf("%d is out of range!\n",s);
    }
    return(0);
}

int verify(int check)
{
    enum { false, true };

    if(check < 0 || check > 100)
        return false;
    return true;
}
```

The code shown in Listing 16-8 is a modification to the solution for Exercise 10-15, from Chapter 10. The *enum* keyword is used at Line 24 to declare two constant values: `false` and `true`. These constants are assigned the values 0 and 1, used in the code to represent TRUE and FALSE conditions.

Exercise 16-9: Type the source code from Listing 16-8 into your editor. Save. Build. Run.

The *enum* keyword is handy for declaring a clutch of sequential constants. In Listing 16-8, it declares `false` and `true`, assigning zero to the value of `false` and 1 to `true`. This keyword saves time over making multiple constant declarations. For example:

```
const int zero = 0;
const int one = 1;
const int two = 2;
const int three = 3;
```

These four statements can be expressed by using a single *enum* statement:

```
enum { zero, one, two, three };
```

The *enum* keyword is followed by a set of curly brackets. They contain the constant declarations, separated by commas, as just shown. The first constant is assigned a value of zero, with each subsequent constant's value incremented by one. This statement can appear in a function, as shown in Listing 16-8, or it can be declared externally.

As a bonus, you can override the default numbering scheme. Consider the following statement:

```
enum { jack=11, queen, king };
```

In this statement, the enumerated constant `jack` is set to a value of 11, `queen` to 12, and `king` to 13. You can use the assignment (=) operator anywhere in the curly brackets to set the value of an enumerated constant.

Exercise 16-10: Write code that asks the user for a number, from 0 to 6, representing a day of the week. Use a *switch-case* structure to evaluate input and output the weekday name.

Here's a sample run of my solution:

```
Enter a weekday number, 0 - 6: 3
Wednesday
```

TIP

You often find enumerated constants used in *switch-case* structures, which helps make the code more readable — if you're clever with the enumerated constant names.

» Showing binary output

» Working with bitwise operators

» Setting and masking bits

» Shifting bits

» Understanding hexadecimal

Chapter 17
Binary Mania

Computers are digital devices, bathed in the binary waters of ones and zeros. Everything your programs deal with — all the text, graphics, music, video, and whatnot — melts down to the basic digital elements of one-zero, true-false, on-off, yes-no. When you understand binary, you can better understand computers and all digital technology.

Binary Basics

Happily, you don't have to program any digital device by writing low-level code, flipping switches, or soldering wires. That's because today's programming happens at a higher level. But still, deep within the machine, that type of low-level coding continues. You're just insulated from the primordial soup of ones and zeros from which all software rises.

Understanding binary

The binary digits, or *bits*, are 1 and 0. Alone, they're feeble; but in groups, they muster great power. Digital storage uses these bits in groups, as illustrated in Table 17-1.

TABLE 17-1 Binary Groupings

Term	C Data Type	Bits	Value Range Unsigned	Value Range Signed
Bit	*_Bool*	1	0 to 1	0 to 1
Byte	*char*	8	0 to 255	−128 to 127
Word	*short int*	16	0 to 65,535	−32,768 to 32,767
Double-word	*Int*	32	0 to 4,294,967,295	−2,147,483,648 to 2,147,483,647
Long	*Long*	64	0 to 18,446,744,073,709,551,615	−9,223,372,036,854,775,807 to 9,223,372,036,854,775,808

The advantage of grouping bits into bytes, words, and so on is that it makes them easier to handle. The processor best deals with information in chunks. How chunks get their values is based upon powers of 2, as shown in Table 17-2.

In Table 17-1, you see the range of values that can be stored in 8 bits, or 1 byte. It's the same range you'd find in a C language *char* variable. Indeed, if you total Column 2 in Table 17-2, you get 255, which is the highest value represented in a byte.

TABLE 17-2 Powers of 2

Expression	Decimal Value	Binary Value
2^0	1	1
2^1	2	10
2^2	4	100
2^3	8	1000
2^4	16	10000
2^5	32	100000
2^6	64	1000000
2^7	128	10000000

TECHNICAL STUFF

A byte has 256 possible values, which includes the all-zero permutation.

Figure 17-1 illustrates how the powers of 2 map into binary storage. Just as decimal places in a base 10 number increase by powers of 10, bits in a binary number increase by powers of 2, reading from right to left.

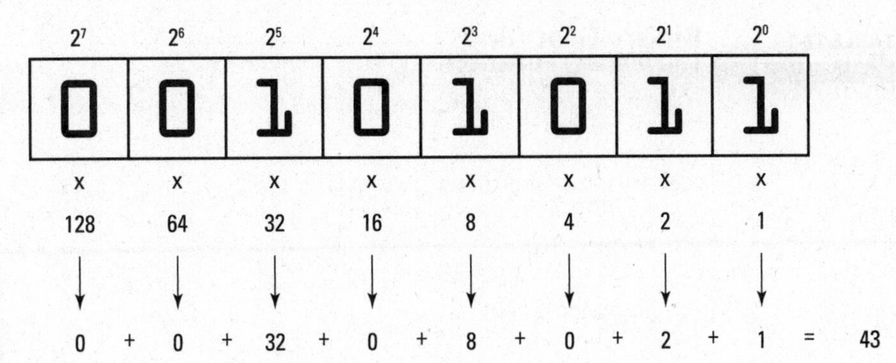

FIGURE 17-1: Base 2 values in a byte.

Each bit that is set or has the value 1 in Figure 17-1 represents a power of two: 2^5, 2^3, 2^1, and 2^0. When you total these values, you get the decimal representation of binary 00101011, which is 43.

That's all well and good, but please don't memorize it!

- Don't concern yourself with translating binary into decimal values; the computer does that job for you all the time. This is because the computer sees only binary and then displays decimal numbers as a courtesy for your human eyeballs. But when you manipulate binary values, it helps to know what's going on.
- Changing a bit's value to 1 is referred to as *setting a bit.*
- Changing a bit's value to 0 is referred to as *resetting a bit.*

Outputting binary values

To best make sense of the C language's binary manipulation operators, it helps to see a binary number in action. The *printf()* function lacks a binary conversion character, and the C library doesn't host a binary output function. Nope, to view a binary number, you must craft your own function.

Listing 17-1 presents a binary output function I've concocted called *binbin()*. The *binbin()* function, at Line 15 in Listing 17-1, swallows an *unsigned char* value. Its output is a string representing that *char* value in binary digits.

LISTING 17-1: The *binbin()* Function

```
#include <stdio.h>

char *binbin(unsigned char n);

int main()
{
    unsigned input;

    printf("Type a value 0 to 255: ");
    scanf("%u",&input);
    printf("%u is binary %s\n",
            input,binbin((unsigned char)input));
    return(0);
}

char *binbin(unsigned char n)
{
    static char bin[9];
    int x;

    for(x=0;x<8;x++)
    {
        bin[x] = n & 0x80 ? '1' : '0';
        n <<= 1;
    }
    bin[x] = '\0';
    return(bin);
}
```

At this point in the chapter, the contents of the *binbin()* function appear rather mysterious. That's okay. The details are offered in the later section "Explaining the binbin() function," and the *char* * thing at the start of the function is discussed in Chapter 19.

Exercise 17-1: Type the source code from Listing 17-1 into your editor. Build and run it a few times to see how integers appear as binary numbers. Try the value 43 to confirm that I got it right in Figure 17-1.

As written in Listing 17-1, *binbin()* displays only 8 bits of data, which is why it's passed an *unsigned char* value — and why the *unsigned* `input` variable is typecast to *unsigned char* in the *printf()* statement (split in the listing at Lines 11 and 2).

Exercise 17-2: Modify the *binbin()* function from Listing 17-1 so that it outputs 16 bits instead of 8. A 16-bit value is a *short int*, and it should be *unsigned* to properly interpret the values. To help you out, change these items:

Line 9: Alter the text so that 65535 is specified instead of 255.

Line 15: Alter the *binbin()* function's argument to be an *unsigned int* (just *unsigned*).

Line 18: Modify the size of the array to 17 to account for 16 characters in the output plus the `\0` (null character) at the end of the string.

Line 21: Adjust the literal value 8 in the code to 16 to account for all 16 characters in the output.

Line 23: Replace the value 0x80 with 0x8000. This change makes the bit field larger, which is something you'll understand better after completing this chapter.

Build Exercise 17-2. Run it a few times to see what the bit field looks like for larger values.

The *binbin()* function, or a variation of it, is used in the following sections to help describe binary programming. You will copy and paste that function frequently, and feel free to use it in your own code however you deem appropriate.

Bit Manipulation

A smattering of C language operators provide data manipulation at the binary level. These operators are easy to ignore, but only when their true power and usefulness aren't appreciated.

Using the bitwise | operator

You're already familiar with the decision-making logical operators: `&&` for AND and `||` for OR. In the `&&` evaluation, both items must be true for the statement to be evaluated as true; for the `||` evaluation, only one of the items must be true.

At the atomic level, the operators `&` and `|` perform similar operations, though on a bit-by-bit basis. The net effect is that you can use the `&` and `|` operators to manipulate individual bits:

The `|` is the bitwise OR operator, also known as the inclusive OR.

The `&` is the bitwise AND operator.

Listing 17-2 demonstrates how to use the bitwise OR operator to set bits in a byte. The OR value is defined as the constant set at Line 7. It's binary 00100000.

LISTING 17-2: The OR Set

```
#include <stdio.h>

char *binbin(unsigned char n);

int main()
{
    const int set = 32;
    unsigned int bor,result;

    printf("Type a value from 0 to 255: ");
    scanf("%u",&bor);
    result = bor | set;

    printf("\t%s\t%d\n",
            binbin((unsigned char)bor),bor);
    printf("|\t%s\t%d\n",
            binbin((unsigned char)set),set);
    printf("=\t%s\t%d\n",
            binbin((unsigned char)result),result);
    return(0);
}

char *binbin(unsigned char n)
{
    static char bin[9];
    int x;

    for(x=0;x<8;x++)
    {
        bin[x] = n & 0x80 ? '1' : '0';
        n <<= 1;
    }
    bin[x] = '\0';
    return(bin);
}
```

Line 12 calculates the bitwise OR operation between a value input, bor, and the set constant. The result is output in three columns: operator, binary string, and decimal value. The result of the operation is that the bits set to 1 in the set value are also set to 1 in the bor value.

Exercise 17-3: Type the source code from Listing 17-2 into your editor to create a new program. Build and run the program.

Here's the output I see for the value 65:

```
Type a value from 0 to 255: 65
          01000001          65
|         00100000          32
=         01100001          97
```

You can see in the binary output how the sixth bit is set in the result.

What does it mean?

The bitwise | (OR) operator lets you manipulate values at the binary level. This control has interesting consequences for certain mathematical operations, as shown in Listing 17-3.

LISTING 17-3: Up with That Text

```
#include <stdio.h>

int main()
{
    char input[64];
    int ch;
    int x = 0;

    printf("Type in ALL CAPS: ");
    fgets(input,63,stdin);

    while(input[x] != '\n')
    {
        ch = input[x] | 32;
        putchar(ch);
        x++;
    }
    putchar('\n');

    return(0);
}
```

Exercise 17-4: Type the source code shown in Listing 17-3 into your editor. Save, build, and run.

Because of the way the ASCII codes map between upper- and lowercase characters, you can switch from upper- to lowercase by setting the sixth bit in a byte.

Using bitwise &

Like the bitwise OR operator, the bitwise AND operator, `&`, also affects bits in a byte. Unlike OR, which sets bits, the AND operation masks bit values. It's easier to show you a program example than to feebly describe what *mask* means.

Exercise 17-5: Modify the source code from Listing 17-2 so that a bitwise AND operation takes place instead of a bitwise OR. Change the constant `set` in Line 7 to the value 223. Change the `|` (bitwise OR) in Line 12 to the `&` (bitwise AND). And finally, change the *printf()* statement in Line 16 so that the `|` is replaced by the `&` character. Build and run.

Here's the output I see when I type the value 255 (all bits set):

```
Type a value from 0 to 255: 255
        11111111        255
&       11011111        223
=       11011111        223
```

The bitwise `&` masks out the sixth bit, causing its value to be reset to 0 in the final calculation. No other bits are affected. To see more examples, try the values 170 and 85. Watch how the bits fall through the mask.

Exercise 17-6: Modify the source code from Listing 17-3 so that a bitwise AND operation takes place instead of a bitwise OR. Change Line 9 so that the *printf()* statement prompts: "Type in some text:" Change Line 14, replacing `|` with `&` and replacing the value 32 with 223. Build and run.

Just as the bitwise OR sets the sixth bit to convert uppercase text to lowercase, masking the sixth bit with a bitwise AND converts lowercase text into uppercase. Of course, the bitwise AND also masks out the space character, changing its value to 0, which isn't a displayable character.

Exercise 17-7: Modify your solution for Exercise 17-6 so that only letters of the alphabet are affected.

Operating exclusively with XOR

XOR is the exclusive OR operator, yet another bitwise logical operator. And to answer your most pressing question, you pronounce XOR like "zor." It's the perfect evil name from bad science fiction.

The XOR operation is kind of weird, but it does have its charm. In this manipulation, bits are compared with one another, just like the & and | operators. When two bits are identical, XOR coughs up a 0. When the two bits are different, XOR spits out a 1. As usual, a program example helps explain things.

The C language XOR operator is the caret character: ^. You can find it put into action on Line 14 in Listing 17-4.

LISTING 17-4: It's Exclusive *or*

```
#include <stdio.h>

char *binbin(unsigned char n);

int main()
{
    int a,x,r;

    a = 73;
    x = 170;

    printf("  %s %3d\n",
            binbin((unsigned char)a),a);
    printf("^ %s %3d\n",
            binbin((unsigned char)x),x);
    r = a ^ x;
    printf("= %s %3d\n",
            binbin((unsigned char)r),r);
    return(0);
}

char *binbin(unsigned char n)
{
    static char bin[9];
    int x;
```

(continued)

LISTING 17-4: *(continued)*

```
    for(x=0;x<8;x++)
    {
        bin[x] = n & 0x80 ? '1' : '0';
        n <<= 1;
    }
    bin[x] = '\0';
    return(bin);
}
```

Exercise 17-8: Type the source code from Listing 17-4 into your editor. Build and run to see how the XOR operation affects binary values.

The charming thing about the XOR operation is that if you use the same XOR value on a variable twice, you get back the variable's original value.

Exercise 17-9: Modify the source code from Listing 17-4 so that one more XOR operation takes place. Insert these three statements after Line 18 (the two *printf()* statements are split between two lines):

```
printf("^ %s %3d\n",
        binbin((unsigned char)x),x);
a = r ^ x;
printf("= %s %3d\n",
        binbin((unsigned char)a),a);
```

Save your update to the code. Build and run. The output looks like this:

```
  01001001  73
^ 10101010 170
= 11100011 227
^ 10101010 170
= 01001001  73
```

Using the same XOR value of 170 turns the value 73 first into 227 and then back to 73.

Because XOR is the exclusive OR operator, some programmers refer to the standard bitwise OR operator as the *inclusive* OR operator.

Understanding the ~ and ! operators

Two infrequent binary operators are the unary ~ (or ones' complement) and the ! (or NOT). They lack the charm of the logical bitwise operators, but I suppose that they have a place.

The ones' complement operator flips all the bits in a value, turning a 1 into a 0 and a 0 into a 1. For example:

```
~01010011 = 10101100
```

The ! (NOT) operator affects the entire value — all the bits. It changes any nonzero value to 0 and the value 0 to 1:

```
!01010011 = 00000000
!00000000 = 00000001
```

REMEMBER

Zero and 1 are the only two results possible when using the bitwise ! operator.

Both the ~ and ! operators are *unary* operators — you prefix a value to get the results.

Table 17-3 summarizes C's binary operators.

TABLE 17-3 Binary Operators

Operator	Name	Type	Action
&	AND	Bitwise	Masks bits, resetting some bits to 0 and leaving the rest alone
|	OR	Bitwise	Sets bits, changing specific bits from 0 to 1
^	XOR	Bitwise	Changes bits to 0 when they match; otherwise, to 1
~	1's complement	Unary	Reverses all bits
!	NOT	Unary	Changes nonzero values to 0; 0 values, to 1

Shifting binary values

The C language features two binary operators that perform the equivalent operation of "Everyone move one step to the left (or right)." The << and >> operators shift bits in value, marching them to the left or right, respectively. Here's the format for the << operator:

```
v = int << count;
```

int is an integer value. *count* is the number of places to shift the value's bits to the left. The result of this operation is stored in variable *v*. Any bits that are shifted to the left beyond the width of the *int* variable *x* are lost. New bits shifted in from the right are always 0.

As with most binary nonsense, it helps to see what's going on in a value when its bits are shifted. Therefore, I present Listing 17-5.

LISTING 17-5: Everyone Out of the Pool!

```
#include <stdio.h>

char *binbin(unsigned char n);

int main()
{
    unsigned bshift,x;

    printf("Type a value 0 to 255: ");
    scanf("%u",&bshift);

    for(x=0;x<8;x++)
    {
        printf("%s\n",
                binbin((unsigned char)bshift));
        bshift = bshift << 1;
    }
    return(0);
}

char *binbin(unsigned char n)
{
    static char bin[9];
    int x;

    for(x=0;x<8;x++)
    {
        bin[x] = n & 0x80 ? '1' : '0';
        n <<= 1;
    }
    bin[x] = '\0';
    return(bin);
}
```

The shift operation takes place at Line 16 in Listing 17-5. The value in variable `bshift` is shifted to the left one bit.

Exercise 17-10: Type the source code from Listing 17-5 into your editor and build a new program to see what it does.

Shifting a value one bit to the left doubles the value. This rule holds true to a certain point: Obviously, the farther left you shift, some bits get lost and the value ceases to double. Also, this trick works best for unsigned values.

Exercise 17-11: Modify the source code from Listing 17-5 so that the *printf()* function at Line 14 also displays the decimal value of the `bshift` variable. You should also modify the *binbin()* function so that it displays 16 digits instead of 8. (Refer to Exercise 17-2 for the 16-bit *binbin()* solution.)

Here's the output I see when using the value 12:

```
Type a value from 0 to 255: 12
0000000000001100 12
0000000000011000 24
0000000000110000 48
0000000001100000 96
0000000011000000 192
0000000110000000 384
0000001100000000 768
0000011000000000 1536
```

Try the value 800,000,000 (don't type the commas) to see how the doubling rule fails as the values keep shifting to the left. Also see the nearby sidebar "Negative binary numbers."

The >> shift operator works similarly to the << shift operator, though values are marched to the right instead of the left. Any bit that's marched off the right end is discarded, and only zero bits slide in on the left. Here's the format:

```
v = int >> count;
```

`int` is an integer value, and *`count`* is the number of places to shift the bits to the right. The result is stored in variable *`v`*.

Exercise 17-12: Modify the source code from Exercise 17-11 so that the right shift operator is used instead of the left shift at Line 15. Build the program.

NEGATIVE BINARY NUMBERS

Binary numbers are always positive, considering that the values of a bit can be only 1 or 0 and not –1 and 0. So how does the computer do signed integers? Easy: It cheats.

The leftmost bit in a signed binary value is known as the *sign bit*. When that bit is set (equal to 1), the value is negative for a *signed int*. Otherwise, the value is read as positive.

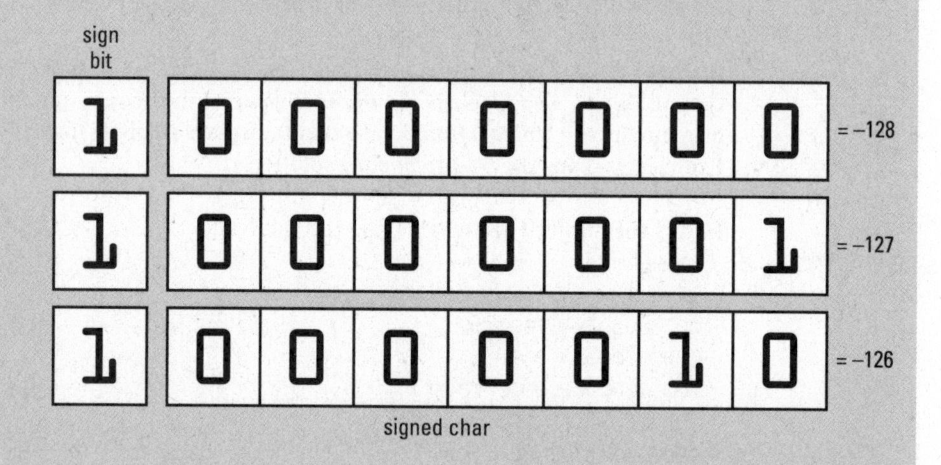

In this example, the sign bit is set for a *signed char*. The values expressed are negative, which is in the range of a *signed char* variable.

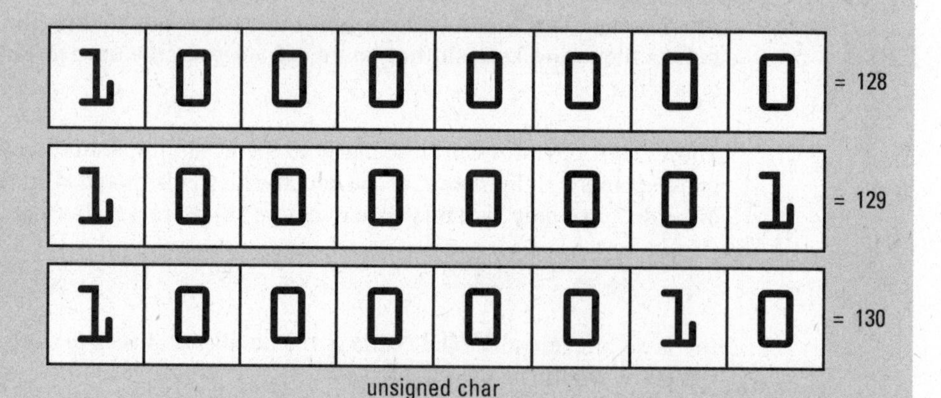

In this example, the sign bit is ignored because the value is an *unsigned char*. The values can only be positive, which is why the positive range for an unsigned variable is greater than for a signed variable.

Here's the result I see when using the value 128:

```
Type a value from 0 to 255: 128
0000000010000000 128
0000000001000000 64
0000000000100000 32
0000000000010000 16
0000000000001000 8
0000000000000100 4
0000000000000010 2
0000000000000001 1
```

TIP

Unlike the << operator, the >> is guaranteed to always cut the value in half when you shift one digit to the right. In fact, the >> operator is far quicker to use on an integer value than the / (division) operator to divide a value by 2.

REMEMBER

Operators looking like the C language's bit shift operators, << and >>, are used in the C++ language for input and output. C++ also has bit shift operators, which are considered overloaded (duplicated) with the I/O operators.

Explaining the binbin() function

If you've worked through this chapter from front to back, I can now sanely explain what's going on in the *binbin()* function to make it convert values into a binary string. Two statements do the job:

```
bin[x] = n & 0x80 ? '1' : '0';
n <<= 1;
```

The first statement performs an AND mask with the value n and hexadecimal value 0x80. All but the leftmost bit in the number is discarded. If that bit is set, which makes it a TRUE condition, the character 1 is stored in the array; otherwise, the character 0 is stored. (Refer to Chapter 8 to review the ternary operator, ?:.)

The value 0x80 is expressed in hexadecimal notation, a type of shorthand for binary. (See the next section, "The Joy of Hex.") The hex value 0x80 is equal to 10000000 binary, which is the AND mask. For wider, 16-bit values, the mask 0x8000 is used instead.

The second statement shifts the bits in the value n one notch to the left. As the loop spins, working through the value n, another bit in the value is shifted to the leftmost position. That bit is evaluated, and the binary string is built by adding a '1' or '0' character to the string.

The Joy of Hex

Face it: No one wants to count bits in a binary number. No one. Perhaps some nerd somewhere can tell you that 10110001 is really the value 177 (I had to look it up), but most programmers can't. What a good programmer can do, however, is translate binary into hex.

Hex has nothing to do with Harry Potter. It's short for *hexadecimal*, which is the base 16 counting system. This concept isn't as obtuse as it sounds, because it's easy to translate between base 16 (hex) and binary.

For example, the value 10110001 translates into B1 hexadecimal. I can see that at once because I've been using hex for a while. It also means that I accept that hexadecimal numbers include the letters *A* through *F*, representing decimal values 10 through 15, respectively. A *B* in hex is the decimal value 11. Letters are used because they occupy a single character position.

Table 17-4 shows the 16 hexadecimal values 0 through F and how they relate to four bits of data.

TABLE 17-4 Hexadecimal Values

Hex	Binary	Decimal	Hex	Binary	Decimal
0x0	0000	0	0x8	1000	8
0x1	0001	1	0x9	1001	9
0x2	0010	2	0xA	1010	10
0x3	0011	3	0xB	1011	11
0x4	0100	4	0xC	1100	12
0x5	0101	5	0xD	1101	13
0x6	0110	6	0xE	1110	14
0x7	0111	7	0xF	1111	15

The hexadecimal values shown in Table 17-4 are prefixed with *0x*. This prefix is used in C, though other programming languages may use different prefixes or a postfix.

The next hexadecimal value after 0xF is 0x10. Don't read it as the number ten, but rather as "one zero hex." It's the value 16 in decimal (base 10). After that, hex keeps counting with 0x11, 0x12, and up through 0x1F and beyond.

Yes, and all of this is just as much fun as learning the ancient Egyptian counting symbols, so where will it get you?

A programmer who sees the binary value 10110100 first splits it into two 4-bit nibbles: 1011 0100. Then he translates it into hex, 0xB4. The C programming language does the translation as well, as long as you use the %x or %X conversion characters, as shown in Listing 17-6.

LISTING 17-6: A Little Hex

```
#include <stdio.h>

char *binbin(unsigned n);

int main()
{
    unsigned b,x;

    b = 21;

    for(x=0;x<8;x++)
    {
        printf("%s 0x%04X %4d\n",binbin(b),b,b);
        b<<=1;
    }

    return(0);
}

char *binbin(unsigned n)
{
    static char bin[17];
    int x;

    for(x=0;x<16;x++)
    {
        bin[x] = n & 0x8000 ? '1' : '0';
        n <<= 1;
    }
    bin[x] = '\0';
    return(bin);
}
```

The code in Listing 17-6 outputs a value in binary, hexadecimal, and decimal and then shifts that value to the left, repeating the process. The *print()* statement at Line 13 uses the `%X` conversion character to output hexadecimal values.

Well, actually, the placeholder is `%04X`, which displays hex values using uppercase letters, four digits wide, and padded with zeros on the left as needed. The `0x` text before the conversion character merely displays the output in standard C style.

Exercise 17-13: Type the code from Listing 17-6 into your editor. Save, build, and run.

Exercise 17-14: Change the value of variable `b` in Line 9 to read this way:

```
b = 0x11;
```

Save this change, build, and run.

TIP

You can write hex values directly in your code. Prefix the values with `0x`, followed by a valid hexadecimal number using either upper- or lowercase letters where required.

TECHNICAL STUFF

ONCE UPON A TIME, OCTAL WAS POPULAR

Another number format available in the C language is octal, or base 8. Octal was quite popular around the time Unix was developed, and many of the old, grizzled programmers league still enjoy tossing around octal values and doing octal-this or octal-that. C even sports an octal conversion character, %o, and an octal prefix, 0 (zero).

I've never used octal in any of my programs. Some older code may use it, and occasionally a function references octal values. So my advice is to be aware of octal, but don't bother to memorize anything.

4
The Advanced Part

IN THIS PART . . .

Discover how variables are stored and accessed

Access variables and memory locations

Replace array notation with pointers

Mangle and abuse an array of pointers

Sort strings by using pointers

Build a linked list of structures

Work with time functions in C

» Grabbing a variable's memory location

» Creating pointer variables

» Peeking at data

» Using pointers to assign values

Chapter 18
Introduction to Pointers

It's considered one of the most frightening topics in all of programming. *Boo!*

Pointers scare a lot of beginning C programmers — and even experienced programmers of other languages. I believe that the reason for the dread is that no one bothers to explain in fun, scintillating detail how pointers really work. So clear your mind, crack your knuckles, and get ready to embrace one of the C language's most unique and powerful features.

The Biggest Problem with Pointers

It's true that you can program in C and avoid pointers. I did it for a long time when I began to learn C programming. Array notation offers a quick-and-dirty workaround for pointers, and you can fake your way around the various pointer functions, hoping that you get it right. But that's not why you bought this book!

After working with pointers for some time and understanding the grief they cause, I've come up with a reason for the woe they induce: Pointers are misnamed.

I can reason why a pointer is called a pointer: It points at something, a location in memory. The problem with this description is that most pedants explain how a pointer works by uttering the phrase, "A pointer points. . . ." This explanation is just wrong. It confuses the issue.

Adding to the name confusion is the fact that pointers have two personalities. One side is a variable that holds a memory location, an *address.* The other side reveals the value at that address. In this way, the pointer should be called a *peeker.* This chapter helps straighten out the confusion.

- The pointer is a part of the C programming language that's considered low-level. It gives you direct access to memory, information that other languages — and even operating systems — prefer that you not touch. For this reason:

WARNING

- A pointer can get you into trouble faster than any other part of C programming. Be prepared to witness memory segmentation errors, bus errors, core dumps, and all sorts of havoc as you experiment with, and begin to understand, pointers.

Sizing Up Variable Storage

Digital storage is measured in bytes. All the information stored inside memory is simply a mass of data, bits piled upon bits, bytes upon bytes. It's up to the software to make sense of it all.

Understanding variable storage

In C, data is categorized by storage type (*char*, *int*, *float*, or *double*) and further classified by keyword (*long*, *short*, *signed*, or *unsigned*). Despite the chaos inside memory, your program's storage is organized into these values, ready for use in your code.

Inside a running program, a variable is described by these attributes:

- Name
- Type
- Value
- Size
- Location

The *name* is the name you give the variable. The name is used only in your code, not when the program runs.

The *type* is one of the C language's data types: *char*, *int*, *float*, or *double*.

The *value* is assigned in your program. Though data at the variable's storage location may exist beforehand, it's considered garbage, and the variable is considered uninitialized until it's assigned a value.

The *size* references the number of bytes of storage the variable occupies.

The *location* is an address, a spot inside the device's memory. This aspect of a variable is something you need not dictate; the program and operating system negotiate where information is stored internally. When the program runs, it uses the location to access a variable's data.

Of these aspects, the variable's name, type, and contents are already known to you. The variable's size and location can also be gathered. Not only that, but the location can be manipulated, which is the inspiration behind pointers.

Reading a variable's size

How big is a *char*? How long is a *long*? You can look up these definitions in Appendix D, but even then the values are general. Only the device you're programming knows the exact storage size of C's standard data types.

Listing 18-1 uses the *sizeof* operator to determine how much storage each C language data type occupies in memory. This operator requires an argument in parentheses and returns a *long unsigned int* value representing the number of bytes the argument — data type, array, structure, and so on — occupies as it squats in memory.

LISTING 18-1: How Big Is a Variable?

```
#include <stdio.h>

int main()
{
    char c = 'c';
    int i = 123;
    long l = 12345678910;
    float f = 98.6;
    double d = 6.022E23;
```

(continued)

LISTING 18-1: *(continued)*

```
    printf("char\t%lu\n",sizeof(c));
    printf("int\t%lu\n",sizeof(i));
    printf("long\t%lu\n",sizeof(l));
    printf("float\t%lu\n",sizeof(f));
    printf("double\t%lu\n",sizeof(d));
    return(0);
}
```

Exercise 18-1: Type the source code from Listing 18-1 into your editor. Build and run to see the size of each variable type.

Here's the output I see:

```
char    1
int     4
long    8
float   4
double  8
```

The value returned by the *sizeof* operator is known as *size_t* data type. Without my getting into a long, boring description, the *size_t* variable is a *typedef* of another variable type, such as a *long unsigned int* on modern computer systems (hence the `%lu` placeholder in Listing 18-1). The bottom line is that the size indicates the number of bytes used to store the operator's argument.

Arrays are also variables in C, and *sizeof* works on them as well, as shown in Listing 18-2.

LISTING 18-2: How Big Is an Array?

```
#include <stdio.h>

int main()
{
    char string[] = "Does this string make me look fat?";

    printf("The string \"%s\" has a size of %u.\n",
            string,sizeof(string));
    return(0);
}
```

Exercise 18-2: Type the source code from Listing 18-2. Build and run it to see how much storage the *char* array occupies.

Exercise 18-3: Edit your source code from Exercise 18-2, adding the *strlen()* function to compare its result on the array with the *sizeof* operator's result.

If the values returned by *strlen()* and *sizeof* differ, can you explain the difference?

Okay, I'll explain: The compiler appends the null character (`\0`) to any string literal, such as the one declared at Line 5 in Listing 18-2. This extra byte of storage is accounted for by the *sizeof* operator. The extra storage is not counted by the *strlen()* function, which returns a character count in the string itself. The terminating null character is a delimiter, not a character in the string.

Exercise 18-4: Edit the source code from Exercise 18-2 again, this time creating an *int* array with five elements. The array need not be assigned any values, nor does it need to be displayed. Build and run.

Can you explain the output? If not, review the output from Exercise 18-1. Try to figure out what's happening.

In Listing 18-3, the *sizeof* operator is used on a structure.

LISTING 18-3: How Large Is a Structure?

```
#include <stdio.h>

int main()
{
    struct robot {
        int alive;
        char name[5];
        int xpos;
        int ypos;
        int strength;
    };

    printf("The evil robot struct size is %lu\n",
            sizeof(struct robot));
    return(0);
}
```

Exercise 18-5: Use Listing 18-3 to create a new source code file. Build and run to determine the size of the structure.

The *sizeof* operator works on all variable types, but for a structure, specify the structure itself. Use the keyword *struct* followed by the structure's name, as shown in Line 14. Avoid using a structure variable when obtaining the size of a structure.

The size of the structure is calculated by totaling the storage requirement for each of its members. You might assume, given the size output from Exercise 18-5, that four *int* variables plus five *char* variables would give you 21: $4 \times 4 + 1 \times 5$. But it doesn't work that way.

On my screen I see this output:

```
The evil robot struct size is 24
```

The reason you see a value other than 21 is that the program aligns variables in memory. It doesn't stack them up, one after another. If I were to guess, I would say that 3 extra bytes are padded to the end of the `name` array to keep it aligned with an 8-byte offset in memory. Figure 18-1 illustrates what's going on.

» The *sizeof* operator returns the size of a C language variable, array, buffer or structure.

» You cannot use *sizeof* to determine the size of your program, the amount of memory in the computer, or the size of anything other than a declared variable or buffer.

REMEMBER

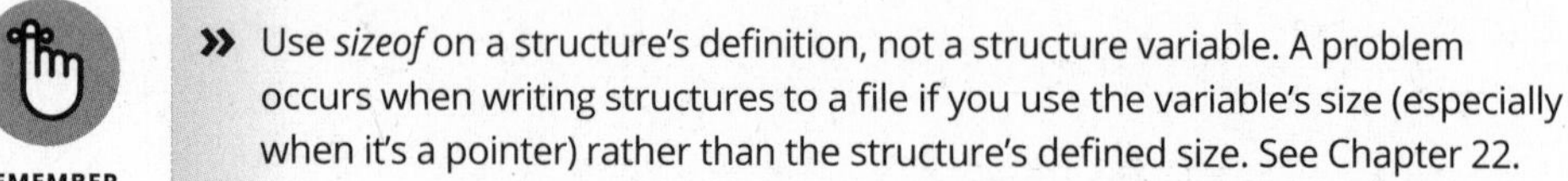

» Use *sizeof* on a structure's definition, not a structure variable. A problem occurs when writing structures to a file if you use the variable's size (especially when it's a pointer) rather than the structure's defined size. See Chapter 22.

» Most compilers today *typedef* the *size_t* value (number of bytes of member) returned by the *sizeof* operator as an *unsigned long* integer, placeholder `%lu`. You can also use the `%zd` placeholder, where `z` represents a byte-size value and `d` stands for decimal output.

» If you're a nerd, you can conclude that the `%zx` placeholder outputs a *size_t* value in hexadecimal.

» The 8-byte offset used to align variables in memory keeps the CPU happy. The processor is much more efficient at reading memory aligned to those 8-byte offsets.

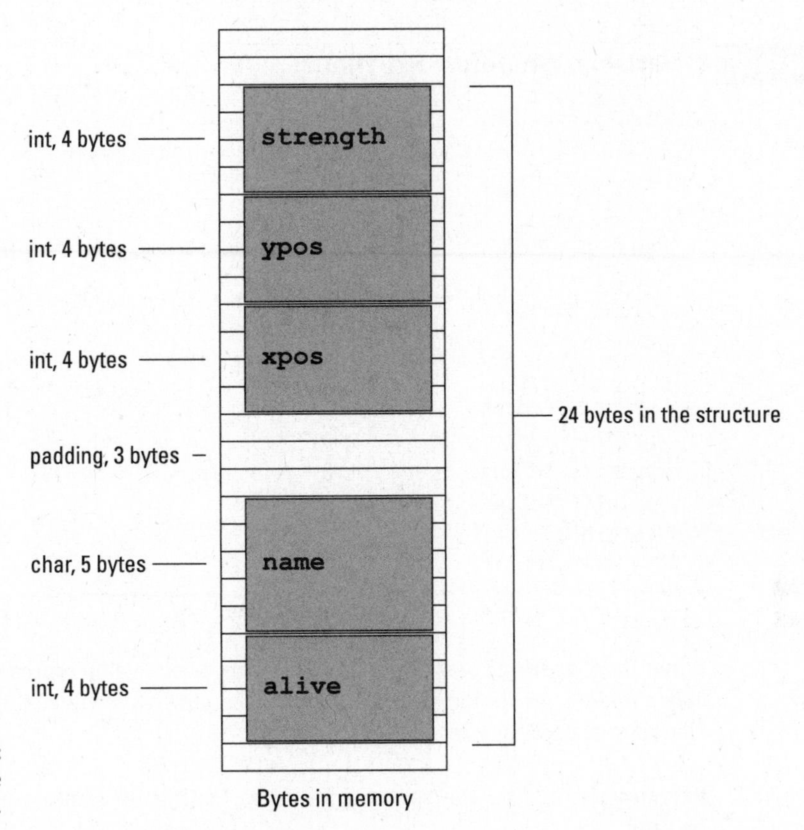

FIGURE 18-1: How a structure fits in memory.

» The values returned by *sizeof* are most likely bytes, as in 8 bits of storage. This size is an assumption: Just about every electronic gizmo today uses an 8-bit byte as the standard storage unit. This assumption doesn't mean you won't find a gizmo with a 7-bit byte or even a 12-bit byte. Just treat the values returned by *sizeof* as a "unit" and you'll be fine.

Checking a variable's location

This chapter has covered four of the attributes used to describe a C language variable: name, type, value, and size. The final description of a variable is its location in memory. You gather this information by using the `&` operator and the `%p` placeholder, as shown in Listing 18-4.

LISTING 18-4: O Variable, Wherefore Art Thou?

```
#include <stdio.h>

int main()
{
    char c = 'c';
    int i = 123;
    float f = 98.6;
    double d = 6.022E23;

    printf("Address of 'c' %p\n",&c);
    printf("Address of 'i' %p\n",&i);
    printf("Address of 'f' %p\n",&f);
    printf("Address of 'd' %p\n",&d);
    return(0);
}
```

When the `&` operator prefixes a variable, it returns a value representing the variable's *address,* or its location in memory. To view this value, the `%p` conversion character is used, as shown in Listing 18-4.

Exercise 18-6: Type the source code from Listing 18-4 into your editor. Build and run.

The results produced by the program generated from Exercise 18-6 are unique, not only for each computer but also, potentially, for each time the program is run. Here's what I see:

```
Address of 'c' 0x7fff5fbff8ff
Address of 'i' 0x7fff5fbff8f8
Address of 'f' 0x7fff5fbff8f4
Address of 'd' 0x7fff5fbff8e8
```

Variable `c` is stored in memory at location 0x7fff5fbff8ff — which is decimal location 140,734,799,804,671. Both values are trivial, of course; the computer keeps track of the memory locations, which is just fine by me. Figure 18-2 offers a memory map of the results just shown.

I can offer no explanation why my computer chose to place the *int* variables where it did, but Figure 18-2 illustrates how those addresses map out in memory.

Individual array elements have memory locations as well, as shown in Listing 18-5 on Line 10. The `&` operator prefixes the specific element variable, coughing up an address. The `%p` conversion character in the *printf()* function outputs the address.

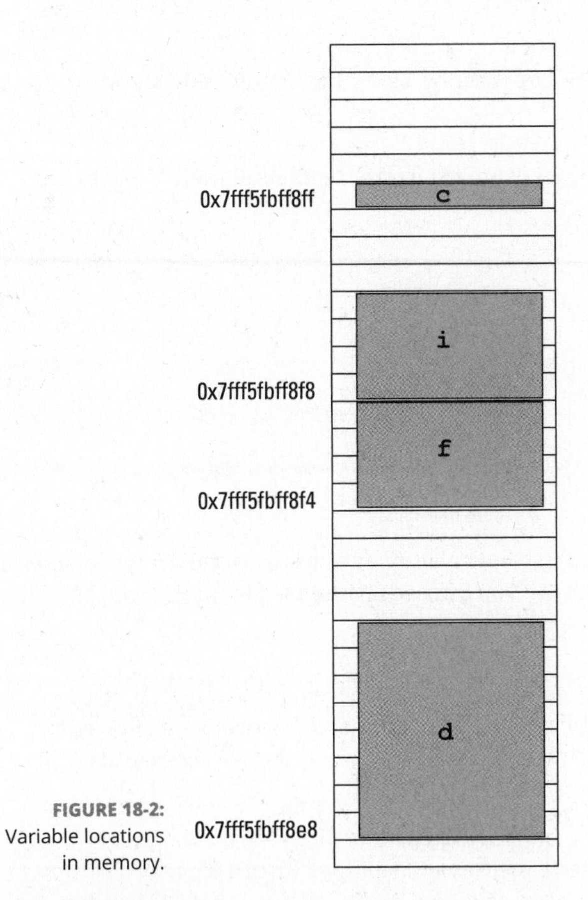

FIGURE 18-2: Variable locations in memory.

LISTING 18-5: Memory Locations in an Array

```
#include <stdio.h>

int main()
{
    char hello[] = "Hello!";
    int i = 0;

    while(hello[i])
    {
        printf("%c at %p\n",hello[i],&hello[i]);
        i++;
    }
    return(0);
}
```

Exercise 18-7: Create a new program by using the source code shown in Listing 18-5. Build and run.

Again, memory location output is unique on each computer. Here's what I see:

```
H at 0x7fff5fbff8f0
e at 0x7fff5fbff8f1
l at 0x7fff5fbff8f2
l at 0x7fff5fbff8f3
o at 0x7fff5fbff8f4
! at 0x7fff5fbff8f5
```

Unlike the example from Exercise 18-6, the addresses generated by Exercise 18-7 are contiguous in memory, one byte after another.

Exercise 18-8: Code a program to display five values in an *int* array along with each element's memory address. You can use Listing 18-5 to inspire you, although a *for* loop might be easier to code.

- By the way, the & address-of operator should be familiar to you. It's used by the *scanf()* function, which requires a variable's address, not the variable itself. The reason is that *scanf()* places a value at a memory location directly. How? By using pointers, of course!
- The & operator is also the bitwise AND operator; however, the compiler is smart enough to tell when & prefixes a variable and when & is part of a binary math equation.

Reviewing variable storage info

To summarize this section, variables in C have a name, type, value, size, and location.

- The variable's type is closely tied to the variable's size in memory, which is obtained by using the *sizeof* operator.
- A variable's value is set or used directly in the code.
- The variable's location is shown courtesy of the & operator and the %p conversion character.

When you have a basic understanding of each of the elements in a variable, you're ready to tackle the hideously complex topic of pointers.

The Hideously Complex Topic of Pointers

Memorize this sentence:

A pointer is a variable that contains a memory location.

Or maybe this story will help:

Once upon a time, a pointer variable met a college student enrolled in a C programming course. The student asked, "What do you point at?" The variable replied, "Nothing! But I contain a memory location." And the freshman was severely satisfied.

You must accept the insanity of the pointer before moving on. True, though you can get at a variable's memory location, or *address*, by using the & operator, the pointer is a far more powerful beast.

Introducing the pointer

A pointer is a type of variable. Like other variables, it must be declared in the code. Further, it must be initialized before it's used. That last part is really important, but first the declaration has this format:

```
type *name;
```

As when you declare any variable, the *type* identifies the pointer as a *char*, *int*, *float*, and so on. The *name* is the pointer variable's name, which must be unique, just like any other variable name. The asterisk identifies the variable as a pointer, not as a regular variable.

The following statement declares a *char* pointer, `sidekick`:

```
char *sidekick;
```

And this statement creates a *double* pointer:

```
double *rainbow;
```

To initialize a pointer, you must assign it a value, just like any other variable. The big difference is that a pointer is initialized to the memory location. This address isn't a random spot in memory, but rather the location of another variable within the program. For example:

```
sidekick = &lead;
```

The preceding statement initializes pointer variable `sidekick` to the address of variable `lead`. Both variables are *char* types; if not, the compiler would growl like a wet cat. After that statement is executed, the `sidekick` pointer contains the address of the `lead` variable.

If you're reading this text and just nodding your head without understanding anything, good! It's time for an example.

I've festooned the source code in Listing 18-6 with comments to help describe two crucial lines. The program really doesn't do anything other than prove that the pointer `sidekick` contains the address, or memory location, of variable `lead`.

LISTING 18-6: An Example

```
#include <stdio.h>

int main()
{
    char lead;
    char *sidekick;

    lead = 'A';          /* initialize char variable */
    sidekick = &lead;    /* initialize pointer IMPORTANT! */

    printf("About variable 'lead':\n");
    printf("Size\t\t%zd\n",sizeof(lead));
    printf("Contents\t%c\n",lead);
    printf("Location\t%p\n",&lead);
    printf("About variable 'sidekick':\n");
    printf("Contents\t%p\n",sidekick);

    return(0);
}
```

Other things to note: Line 12 uses two tab escape sequences to line up the output. Also, don't forget the `&` in Line 14, which fetches the variable's address.

Exercise 18-9: Type the source code from Listing 18-6 into your editor. Build and run.

Here's the output I see on my screen:

```
About variable 'lead':
Size            1
Contents        A
Location        0x7fff5fbff8ff
About variable 'sidekick':
Contents        0x7fff5fbff8ff
```

The addresses (in the example) are unique for each system, but the key thing to note is that the contents of pointer `sidekick` are equal to the memory location of variable `lead`. That's because of the initialization that takes place on Line 9 in the code:

```
sidekick = &lead;
```

It would be pointless for a pointer to merely contain a memory address. The pointer can also peek into its address and determine the value that's stored there. To make that happen, the * operator is prefixed to the pointer's variable name.

Exercise 18-10: Modify your source code from Exercise 18-9 by adding the following statement after Line 16:

```
printf("Peek value\t%c\n",*sidekick);
```

Build and run. Here's the output I see as output:

```
About variable 'lead':
Size            1
Contents        A
Location        0x7fff5fbff8ff
And variable 'sidekick':
Contents        0x7fff5fbff8ff
Peek value      A
```

When you specify the * (asterisk) before an initialized pointer variable's name, the results are the contents of the address. The value is interpreted based on the type of pointer. In this example, `*sidekick` represents the *char* value stored at a memory location kept in the `sidekick` variable, which is really the same as the memory location variable `lead`.

To put it another way:

- A pointer variable contains a memory location.
- The *pointer variable peeks into the value stored at that memory location.

Working with pointers

The pointer's power comes from its split personality as well as from its capability to manipulate values at its stored memory location.

In Listing 18-7, three *char* variables are declared at Line 5 and initialized all on Line 8. (I stacked them up on a single line so that the listing wouldn't get too long.) A *char* pointer is created at Line 6 and then initialized at Lines 11, 13, and 15.

LISTING 18-7: More Pointer Fun

```
#include <stdio.h>

int main()
{
    char a,b,c;
    char *p;

    a = 'A'; b = 'B'; c = 'C';

    printf("Know your ");
    p = &a;                     /* initialize */
    putchar(*p);                /*  use */
    p = &b;                     /* initialize */
    putchar(*p);                /*  use */
    p = &c;                     /* initialize */
    putchar(*p);                /*  use */
    printf("s\n");

    return(0);
}
```

Lines 11 and 12 set up the basic operation in the code: First, pointer p is initialized to the address of a *char* variable. Second, the * (asterisk) peeks at the value stored at that address. The *p variable represents that value as a *char* inside the *putchar()* function. This operation is then repeated for *char* variables b and c.

Exercise 18-11: Create a new project by using the source code from Listing 18-7. Build and run.

Figure 18-3 attempts to illustrate the behavior of pointer variable p as the code runs.

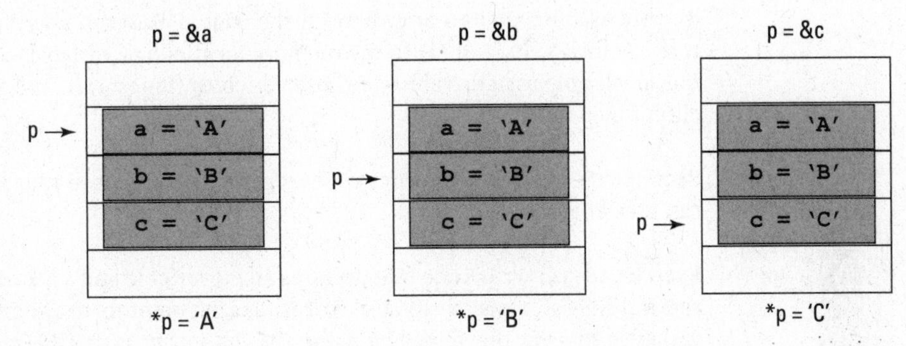

FIGURE 18-3: Using a pointer to read values.

Exercise 18-12: Write a program that declares both an *int* variable and an *int* pointer variable. Use the pointer variable to display the value stored by the *int* variable.

Just as you can grab a variable's value, as shown in Listing 18-7, you can use the *pointer operator to set a variable's value. Refer to Listing 18-8.

LISTING 18-8: Assigning Values by Using a Pointer

```
#include <stdio.h>

int main()
{
    char a,b,c;
    char *p;

    p = &a;             /* initialize */
    *p = 'A';           /*  assign */
    p = &b;             /* initialize */
    *p = 'B';           /*  assign */
```

(continued)

LISTING 18-8: *(continued)*

```
    p = &c;                  /* initialize */
    *p = 'C';                /*  assign */
    printf("Know your %c%c%cs\n",a,b,c);
    return(0);
}
```

Line 5 in Listing 18-8 declares three *char* variables. These variables are never directly assigned values anywhere in the code. Pointer p, however, is initialized thrice (Lines 8, 10, and 12) to the memory locations of variables a, b, and c. Then the *p variable assigns values to those variables (Lines 9, 11, and 13.) The result is output by *printf()* at Line 14.

Exercise 18-13: Copy the source code from Listing 18-8 into your editor. Build and run the program.

Exercise 18-14: Write code that declares an *int* variable and a *float* variable for your age and weight, respectively. Use corresponding pointers to assign values to these variables. Output the values by using the *int* and *float* variables, not the pointer variables.

» Replacing array notation with pointers

» Working with strings and pointers

» Understanding arrays of pointers

» Performing a string sort

» Creating a function that eats pointers

Chapter 19
Deep into Pointer Land

It's easy to accept what a pointer does, to numbly nod your head, to repeat the mantra, "A pointer is a variable that contains a memory location." You can even memorize the difference between pointer variable p and pointer variable *p. But to truly know the power of the pointer, you have to discover how it's fully exploited in the C language. You must eschew the old way of doing things and fully embrace pointers for the miraculous witchcraft they do.

Pointers and Arrays

Arrays in the C language are nothing but a kettle full of lies! Truly, they don't exist. As you discover the power of the pointer, you come to accept that an array is merely a cleverly disguised pointer. Be prepared to feel betrayed.

Getting the address of an array

An array is a type of variable in C, one that you can examine for its size and address. Chapter 18 covers using the *sizeof* operator on an array. Now you uncover the deep, dark secret of beholding an array's address.

The source code from Listing 19-1 shows a teensy program that declares an *int* array and then displays that array's location in memory. Simple. (Well, it's simple if you've worked through Chapter 18.)

LISTING 19-1: Where the Array Lurks

```
#include <stdio.h>

int main()
{
    int array[5] = { 2, 3, 5, 7, 11 };

    printf("'array' is at address %p\n",&array);
    return(0);
}
```

Exercise 19-1: Type the source code from Listing 19-1 into your editor. Build and run the program.

Here's the output I see:

```
'array' is at address 0x7fffcb7333b0
```

Exercise 19-2: Duplicate Line 7 in the code to create a new Line 8, removing the ampersand:

```
printf("'array' is at address %p\n",array);
```

The difference between the old Line 7 and the new Line 8 is the missing `&` that prefixes the `array` variable. Will it work? Compile and run to be sure.

Here's my output for the new code:

```
'array' is at address 0x7fffeddd9c40
'array' is at address 0x7fffeddd9c40
```

Is the `&` prefix necessary? Better find out:

Exercise 19-3: Summon the source code from Exercise 18-6 (from Chapter 18). Edit Lines 10 through 14 to remove the `&` from the variable's name in the *printf()* statement. Attempt to build the program.

Here's the warning message I saw repeated four times:

```
Warning: format specifies type 'void *' ...
```

Obviously, the & is important for individual variables. But for arrays, it's optional. But how could that be, unless . . . unless an array is really a pointer!

Working pointer math in an array

What happens when you increment a pointer? Say that pointer variable dave references a variable at memory address 0x8000. If so, consider this statement:

```
dave++;
```

What would the new value of pointer dave be?

Your first inclination might be to say that dave would be incremented by 1, which is correct. But the result of the calculation may not be 0x8001. That's because the address stored in a pointer variable is incremented by one *unit*, not by one digit.

What's a unit?

It depends on the pointer's data type. If pointer dave is a *char* pointer, indeed the new address could be 0x8001. But if dave were an *int* or a *float*, the new address would be the same as

```
0x8000 + sizeof(int)
```

or

```
0x8000 + sizeof(float)
```

On most systems, an *int* is 4 bytes, so you could guess that dave would be 0x8004 after the increment operation. But why guess when you can code?

Listing 19-2 illustrates a simple program, something I could have directed you to code without using pointers: Fill an *int* array with values 1 through 10, and then display the array's elements and their values. But in Listing 19-2, a pointer is used to fill the array.

LISTING 19-2: Arrays and Pointer Math

```
#include <stdio.h>

int main()
{
    int numbers[10];
    int x;
    int *pn;

    pn = numbers;        /* initialize pointer */

/* Fill array */
    for(x=0;x<10;x++)
    {
        *pn=x+1;
        pn++;
    }

/* Display array */
    for(x=0;x<10;x++)
        printf("numbers[%d] = %d\n",
                x,numbers[x]);

    return(0);
}
```

Line 7 declares the pointer pn, and Line 9 initializes it. The & isn't needed here, because numbers is an array, not an individual variable. At that point, the pointer holds the base address of the array, as illustrated in Figure 19-1. Keep in mind that the array is empty.

The *for* loop at Lines 12 through 16 fills the numbers array. The first element is filled at Line 14 using the peeker notation for pointer pn. Then at Line 15, pointer pn is incremented one unit. It now points at the next element in the array, as shown in Figure 19-1, and the loop repeats.

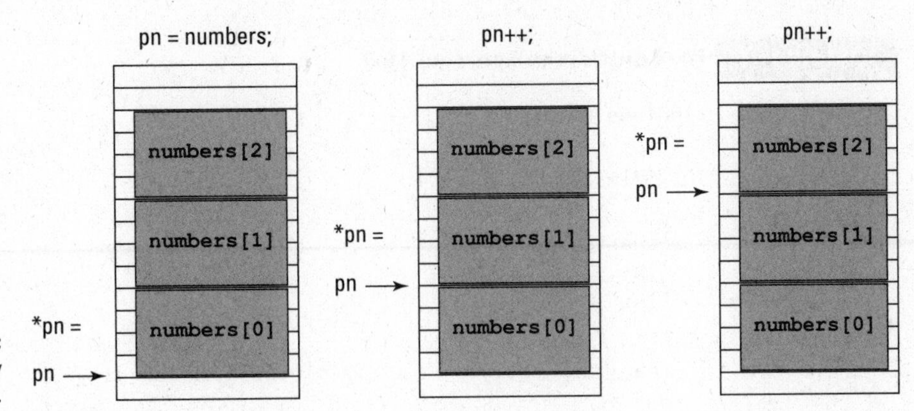

FIGURE 19-1: Filling an array by using a pointer.

Exercise 19-4: Copy the source code from Listing 19-2 into your editor. Build and run.

Exercise 19-5: Modify your source code from Exercise 19-4 so that the address of each element in the array is displayed along with its value.

In the output of Exercise 19-5, you should see that each address is separated by 4 bytes (assuming that the size of an *int* is 4 bytes on your machine). In fact, the addresses probably all end in the hex digits 0, 4, 8, and C.

Exercise 19-6: Complete the conversion of Listing 19-2, and what you began in Exercise 19-5, by having the second *for* loop display the array's values using the peeker side of pointer variable pn.

Exercise 19-7: Create a new project that fills a 26-character array by using pointers similar to the ones shown in Listing 19-2. Fill the array with the letters 'A' through 'Z' by using pointer notation. (Build a *char* array, not a string.) Use pointer notation to output the array's contents.

Here's a big hint:

```
*pn=x+'A';
```

In fact, in case you're totally lost, I've put my solution for Exercise 19-7 in Listing 19-3.

LISTING 19-3: My Solution to Exercise 19-7

```
#include <stdio.h>

int main()
{
    char alphabet[26];
    int x;
    char *pa;

    pa = alphabet;              /* initialize pointer */

    /* fill the array */
    for(x=0;x<26;x++)
    {
        *pa=x+'A';
        pa++;
    }

    pa = alphabet;              /* re-initialize pointer */

    /* output the array */
    for(x=0;x<26;x++)
    {
        putchar(*pa);
        pa++;
    }
    putchar('\n');

    return(0);
}
```

The source code in Listing 19-3 should be rather lucid, performing each task one step at a time. But keep in mind that many C programmers like to combine statements, and such combinations happen frequently with pointers.

Exercise 19-8: Combine the two statements in the first *for* loop from Listing 19-3 to be only one statement:

```
*pa++=x+'A';
```

Ensure that you type it in properly. Save, build, and run.

The output is the same. What this ugly `*pa++=x+'A'` mess does is described here:

x+'A' This part of the statement is executed first, adding the value of variable x to letter A. The effect is that the code marches up the alphabet as the value of x increases.

*pa The result of `x+'A'` is placed into the memory location specified by pointer pa.

++ The value of variable pa — the memory address — is incremented one unit. Because the ++ appears after the variable (it's post-fixed), the value is incremented *after* the value at that address is written.

Keeping the two statements separate still works, and I code my programs this way because it's easier for me to read later. But not every programmer does so! Many of them love to stack up pointers with the increment operator. Watch out for it!

Exercise 19-9: Fix up your source code from Exercise 19-8 so that the second *for* loop uses the `*pa++` monster.

Hopefully, the `*pa++` pointer-thing makes sense. If not, take a nap and then come back and examine Listing 19-4.

LISTING 19-4: Head-Imploding Program

```
#include <stdio.h>

int main()
{
    char alpha = 'A';
    int x;
    char *pa;

    pa = <SPiSymbol>&alpha;</SPiSymbol>          /* initialize
  pointer */

    for(x=0;x<26;x++)
        putchar((*pa)++);
    putchar('\n');

    return(0);
}
```

The source code from Listing 19-4 deals with a single *char* variable and not an array. Therefore, the pointer initialization in Line 9 requires the & prefix. Don't forget it!

Line 12 in this code contains the booger (*pa)++. It looks similar to *pa++, but it's definitely not. Unlike *pa++, which peeks at a value and then increments the pointer, the (*pa)++ construction increments a value being peeked at; the pointer's address is unchanged.

Exercise 19-10: Edit, build, and run a new program by using the source code from Listing 19-4.

The (*pa)++ operation works, thanks to the parentheses. The program fetches the value represented by *pa first, and then that value is incremented. The pointer variable, pa, isn't affected by the operation; pa still holds the address of *char* variable alpha.

To help avoid confusion on this topic, I offer Table 19-1, which explains the various cryptic pointer/peeker notation doodads.

TABLE 19-1 Pointers and Peekers In and Out of Parentheses

Expression	Address p	Value *p
*p++	Incremented after the value is read	Unchanged
*(p++)	Incremented after the value is read	Unchanged
(*p)++	Unchanged	Incremented after it's read
*++p	Incremented before the value is read	Unchanged
*(++p)	Incremented before the value is read	Unchanged
++*p	Unchanged	Incremented before it's read
++(*p)	Unchanged	Incremented before it's read

Use Table 19-1 to help you decipher code as well as get the correct format for what you need done with a pointer. If the pointer notation you see or want doesn't appear in Table 19-1, it's either not possible or not a pointer. For example, the expressions p*++ and p++* may look like they belong in Table 19-1, but they're not pointers. (In fact, they're not defined as valid expressions in C.)

Substituting pointers for array notation

Array notation must be a myth because it can easily be replaced by pointer notation. In fact, internally to your programs, it probably is.

Consider Table 19-2, which compares array notation with pointer notation. Assume that pointer `a` is initialized to array `alpha`. The array and pointer are of the same variable type; notation doesn't differ between variable types. A *char* array and an *int* array would use the same references, as shown in both columns in Table 19-2.

TABLE 19-2 Array Notation Replaced by Pointers

Array alpha[]	*Pointer* a
alpha[0]	*a *or* *(a+0)
alpha[1]	*(a+1)
alpha[2]	*(a+2)
alpha[3]	*(a+3)
alpha[*n*]	*(a+*n*)

You can test your knowledge of array-to-pointer notation by using a sample program, such as the one shown in Listing 19-5.

LISTING 19-5: A Simple Array Program

```
#include <stdio.h>

int main()
{
    enum weekdays { mon, tues, wed, thurs, fri };
    float temps[5] = { 18.7, 22.8, 25.0, 23.3, 23.2 };

    printf("The temperature on Tuesday was %.1f\n",
            temps[tues]);
    printf("The temperature on Friday was %.1f\n",
            temps[fri]);
    return(0);
}
```

At Line 5, the *enum* keyword creates a list of enumerated constants to represent values 0 through 4. Each constant refers to an element in the `temps` array, corresponding to a day of the week. This example shows how enumerated constants are used to make a program more readable.

Exercise 19-11: Rewrite the code so that the two *printf()* statements from Listing 19-5 use pointer notation to output the values. You may use literal values or the enumerated constants in your code.

Strings Are Pointer-Things

C lacks a string data type, but it does have the *char* array, which is effectively the same thing. As an array, a string in C can be completely twisted, torqued, and abused by using pointers. It's a much more interesting topic than messing with numeric arrays, which is why it gets a whole section all by itself.

Using pointers to display a string

You're most likely familiar with displaying a string in C, probably by using either the *puts()* or *printf()* function. Strings, too, can be displayed one character a time by plodding through an array. To wit, I offer Listing 19-6.

LISTING 19-6: Hello, String

```
#include <stdio.h>

int main()
{
    char sample[] = "From whence cometh my help?\n";
    int index = 0;

    while(sample[index] != '\0')
    {
        putchar(sample[index]);
        index++;
    }
    return(0);
}
```

The code shown in Listing 19-6 is completely legitimate C code, valid to create a program that displays a string. But it doesn't use pointers, does it?

Exercise 19-12: Modify the source code from Listing 19-6, replacing array notation with pointer notation. Eliminate the `index` variable. You need to create and initialize a pointer variable.

It's possible to tighten the *while* loop's evaluation in Listing 19-6. The null character evaluates as FALSE. So the evaluation could be rewritten as

```
while(sample[index])
```

The loop spins as long as the array element referenced by `sample[index]` isn't a null character.

Exercise 19-13: Edit the *while* loop's evaluation in your solution for Exercise 19-12, eliminating the null character comparison.

Exercise 19-14: Continue working on your code, this time eliminating all statements in the *while* loop. Set all the action in the *while* statement's evaluation. For the sake of reference, the *putchar()* function returns the character that's output.

Using a pointer to declare a string

Here's a scary trick you can pull using pointers, one that comes with a boatload of caution. Consider Listing 19-7.

LISTING 19-7: A Pointer Announces a String

```
#include <stdio.h>

int main()
{
    const char *sample = "From whence cometh my help?\n";

    puts(sample);
    return(0);
}
```

In Listing 19-7, the string output is created by initializing a pointer. It's a construct that looks odd, but it's something you witness often in C code, particularly with strings. (You cannot use this convention to initialize a numeric array.)

Exercise 19-15: Copy the source code from Listing 19-7 in your editor. Build and run.

Here is the boatload of caution: First, the string is declared as a constant, which I strongly recommend. The reason is that this construction doesn't behave the same way as a traditional string declaration and manipulating the string must be avoided.

Second, the pointer value (its address) shouldn't be changed (incremented, for example,) because doing so loses the string's location in memory.

Third, processing this type of string declaration may result in unintended and unpredictable consequences. For example, if you use the *putchar()* function with a pointer to output the string, the null character may not be interpreted properly.

Fourth, this type of string has issues when passed to a function, which is why I offer this warning:

WARNING

When declaring a string by using a pointer, don't mess with the pointer variable elsewhere in the code.

The solution is to declare strings as arrays and just leave it at that.

Building an array of pointers

An array of pointers would be an array that holds memory locations. Such a construction is often necessary in C, and I could devise a wickedly complex demo program that would frustrate you to insanity. But this condition doesn't result when you consider that an array of pointers is really an array of strings, as shown in Listing 19-8. This approach makes topic digestion easier.

LISTING 19-8: Crazy Pointer Arrays

```
#include <stdio.h>

int main()
{
    char *fruit[] = {
        "watermelon",
        "banana",
```

```
        "pear",
        "apple",
        "coconut",
        "grape",
        "blueberry"
    };
    int x;

    for(x=0;x<7;x++)
        puts(fruit[x]);

    return(0);
}
```

An array of pointers is declared in Listing 19-8. It works similarly to Listing 12-7 (from Chapter 12), though in this construction you don't need to specifically count individual string lengths. That's because the array is really an array of pointers, or memory locations. Each string dwells somewhere in memory. The array lists where each one starts.

Exercise 19-16: Type the source code from Listing 19-8 into your editor. Build and run to confirm that it works.

This chapter covers pointers, so which part of Listing 19-8 do you think could be improved?

Exercise 19-17: Using information from Table 19-2 as your guide, replace the array notation at Line 17 in Listing 19-8 with pointer notation.

The reason that your solution to Exercise 19-17 works (assuming that you got it correct) is that the fruit array contains pointers. The value of each element is another pointer. But that's nothing; consider Listing 19-9.

LISTING 19-9: Pointers-to-Pointers Example

```
#include <stdio.h>

int main()
{
    char *fruit[] = {
        "watermelon",
        "banana",
```

(continued)

LISTING 19-9: (continued)

```
        "pear",
        "apple",
        "coconut",
        "grape",
        "blueberry"
    };
    int x;

    for(x=0;x<7;x++)
    {
        putchar(**(fruit+x));
        putchar('\n');
    }

    return(0);
}
```

Line 18 in Listing 19-9 contains the dreaded, feared, avoided, and cursed `**` notation, or *double-pointer* notation. To use my preferred nomenclature, it's a double-peeker. Before I commence the discussion, do Exercise 19-18.

Exercise 19-18: Carefully type the source code from Listing 19-9 into your editor. Compile and run.

To understand the `**(fruit+x)` construct, you must work from the inside out:

```
fruit+x
```

Variable `fruit` contains a memory address. It's a pointer! The `x` is a value incrementing by one unit. In this case, the unit is the size of a pointer; all elements of the `fruit` array are pointers.

```
*(fruit+x)
```

You've seen the preceding construction already. It's the contents of the address `fruit+x`. From the code, `fruit` is an array of pointers. So the result of the preceding operation is . . . a pointer!

```
**(fruit+x)
```

Finally, you get a pointer to a pointer or — put better — a peeker to a peeker. If the inside peeker is a memory address, the outside peeker (the first asterisk) is the content of that memory address. Figure 19-2 attempts to clear up this concept.

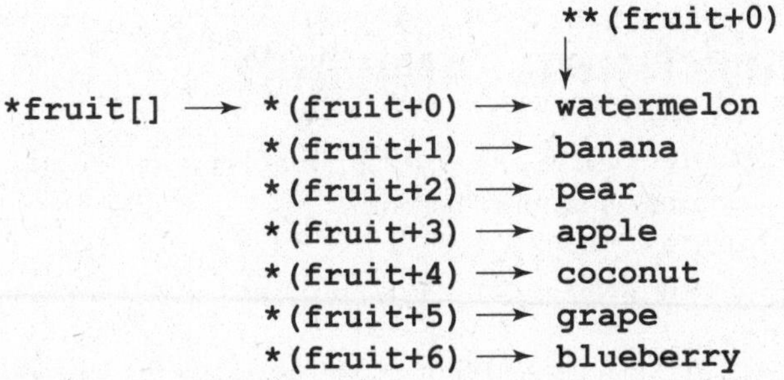

```
                                  **(fruit+0)
                                       |
                                       v
*fruit[] --> *(fruit+0) --> watermelon
             *(fruit+1) --> banana
             *(fruit+2) --> pear
             *(fruit+3) --> apple
             *(fruit+4) --> coconut
             *(fruit+5) --> grape
             *(fruit+6) --> blueberry
```

FIGURE 19-2: How the ** thing works.

It helps to remember that the ** operator is almost always (but not exclusively) tied to an array of pointers; or, if you want to make it simple, to an array of strings. So, in Figure 19-2, *fruit[] represents the address of an array of pointers. The second column contains each address as it's referenced relative to the fruit base address. The column on the right shows the strings at each address, with the **(fruit+0) expression holding the first character of the first string. Subsequent strings' first characters would be **(fruit+1), **(fruit+1), and so on.

If you're still confused — and I don't blame you; Einstein was in knots at this point when he read this book's first edition — consider mulling over Table 19-3. In the table, pointer notation (using variable ptr) is compared with the equivalent array notation (using variable array).

TABLE 19-3 Pointer Notation and Array Notation

Pointer Notation	Array Notation	Description
**ptr	*array[]	Declares an array of pointers
*ptr	array[0]	The address of the first pointer in the array; for a string array, the first string
*(ptr+0)	array[0]	The same as the preceding entry
**ptr	array[0][0]	The first element of the first pointer in the array; the first character of the first string in the array
**(ptr+1)	array[1][0]	The first element of the second pointer in the array; the first character of the second string
((ptr+1))	array[1][0]	The same as the preceding entry
((ptr+a)+b)	array[a][b]	Element b of pointer a

SEE ANYTHING FAMILIAR?

Arrays of pointers should be somewhat familiar to you. If you've worked through Chapter 15, you may remember this thing:

```
int main(int argc, char *argv[])
```

The full declaration of the *main()* function includes an array of pointers as the second argument. In Chapter 15, you treat each item as its own string, which is exactly how the construction works. But secretly, what's being passed to the *main()* function is an array of pointers.

By the way, the full declaration of the *main()* function can also be written like this:

```
int main(int argc, char **argv)
```

Just to keep you from understanding this topic too well, accept that the `*array[]` notation can also be written as `**array`. Effectively, the `**` declaration refers to a pointer to a pointer.

Exercise 19-19: Rework your source code from Exercise 19-18 so that each individual character in a string is displayed, one at a time, by using the *putchar()* function. If you can write the entire *putchar()* operation as a *while* loop's condition, you get ten *For Dummies* bonus points.

Sorting strings

Taking what you know about sorting in the C language (gleaned from Chapter 12), you can probably craft a decent string sorting program. Or, at minimum, you can explain how it's done. That's great! But it's a lot of work.

What's better when it comes to sorting strings is not to sort the strings at all. No, instead, you sort an array of pointers referencing the strings. Listing 19-10 shows an example.

LISTING 19-10: Sorting Strings, Initial Attempt

```
#include <stdio.h>

int main()
{
```

```
    char *fruit[] = {
        "apricot",
        "banana",
        "pineapple",
        "apple",
        "persimmon",
        "pear",
        "blueberry"
    };
    char *temp;
    int a,b,x;

    for(a=0;a<6;a++)
        for(b=a+1;b<7;b++)
            if(*(fruit+a) > *(fruit+b))
            {
                temp = *(fruit+a);
                *(fruit+a) = *(fruit+b);
                *(fruit+b) = temp;
            }

    for(x=0;x<7;x++)
        puts(fruit[x]);

    return(0);
}
```

Exercise 19-20: Type the source code from Listing 19-10 into your editor. Build and run to ensure that the strings are properly sorted.

Well, it probably didn't work. It may have, but if the list is sorted or changed in any way, it's an unintended consequence and definitely not repeatable.

The problem is in Line 19: You can't compare strings by using the > operator. You can compare individual characters and then sort the list based on those characters, but most humans prefer words sorted across their entire length, not just the first character.

Exercise 19-21: Modify your source code, and use the *strcmp()* function to compare strings to determine whether they need to be swapped.

Pointers in Functions

A pointer is a type of variable. As such, it can easily be flung off to a function. Even more thrilling, a pointer can wander back from a function as a return value. Oftentimes, these tricks are the only ways to get information to or from a function.

Passing a pointer to a function

The great advantage of passing a pointer to a function is that the information that's modified is automatically returned. That's because the function references a memory address, not a value directly. By using that address, information can be manipulated directly. Listing 19-11 demonstrates.

LISTING 19-11: Pointing at a Discount

```
#include <stdio.h>

void discount(float *a);

int main()
{
    float price = 42.99;

    printf("The item costs $%.2f\n",price);
    discount(&price);
    printf("With the discount, that's $%.2f\n",price);
    return(0);
}

void discount(float *a)
{
    *a *= 0.90;
}
```

In Line 3 of Listing 19-11, the *discount()* function is prototyped. It requires a *float* pointer variable as its only argument.

Line 10 passes the address of the `price` variable to the *discount()* function. The & operator obtains the memory location of the `price` variable.

Within the function, pointer variable `a` accesses the value stored at its memory location. An assignment operator reduces the value by 90 percent. Nothing is returned, because the value is modified directly in memory.

Exercise 19-22: Type the source code from Listing 19-11 into your editor. Build and run the program.

Exercise 19-23: Write code in which the *swap()* function swaps the values of two integer variables. In the *main()* function, output the variable's values before and after the swap. The *swap()* function returns no values; it's of the *void* data type.

Returning a pointer from a function

Functions are known by their types, such as *int* or *char* or even *void*. You can also declare pointer functions, which return a memory location as a value. For example:

```
int *monster(void);
```

In this example, the *monster()* function requires no arguments but returns a pointer to an integer value.

Most functions that return pointers return the memory location of a buffer — a storage area that the function allocates. For example:

```
struct person *fill_data(void);
```

The preceding prototype declares the *fill_data()* function, which returns the address of a `person` structure. The structure is created, or allocated, within the function and its memory address returned.

The *address()* function prototype shown on the following line returns a string value, either a *static* array declared in the function or the address of a freshly allocated *char* buffer:

```
char *address(int a, int b);
```

Allocating memory buffers is the job of the *malloc()* function. This function, as well as many examples of functions that return memory locations, is covered in Chapter 20.

» Initializing and allocating memory

» Resizing an allocated memory chunk

» De-allocating memory

» Creating space for a structure

» Building a linked list

» Editing structures in a linked list

Chapter 20

Memory Chunks and Linked Lists

Another reason for having a pointer variable is to hold the address of a freshly allocated chunk of memory. This approach is far better than creating an array and guessing at its size: Memory chunks can be assigned a size on the fly, resized, and banished. You can't perform such actions with an array — well, not in civilized society.

Dovetailing from memory allocation comes the ultimate thrill ride in the C language amusement park: the linked list. It combines the mystery of structures with the dread of pointers to create one heart-stopping, scream-inducing, hair-blowing roller coaster of fun. Please fasten your gaming chair's seat belt and keep your hands on the keyboard at all times. Here we go!

Give Me Memory!

Don't tell a beginning programmer, but declaring a variable in C is in reality directing the program to beg for some storage space from the operating system. As you know (hopefully, you know), the operating system is the Lord High Master of the computer or whatever electronic device you're programming. As such, it doles out RAM to programs that request it.

When you declare a variable, from a lowly *short int* to a massive string buffer, you're directing the program to beg for that much space, into which you plan to put something useful. In the C language, you can also allocate memory on the fly, as long as you have an army of pointers at hand to save the addresses.

Introducing the malloc() function

The *malloc()* function allocates a chunk of memory; think "memory allocation," malloc. You pass it a *size_t* (byte) value, and it returns the address — a pointer — for the allocated chunk. When something horrible happens, it returns the NULL pointer constant. Here's the format:

```
p = malloc(size);
```

The `size` argument is a *size_t* value, the same data type returned by the *sizeof* operator. Unofficially, it's a byte. Officially, programmers set this value based on the size of the data type they want to store in the memory chunk. More on that in a bit.

The value returned is a pointer, `p`, a memory location. The NULL constant is returned otherwise, which must be checked for to confirm that memory was allocated. If not, your program gets into all kinds of trouble.

TECHNICAL STUFF

The NULL constant is not the same thing as the null character, `\0`, marking the end of a string. NULL is a pointer equivalent used to test for unallocated memory. The null character is character code zero, a byte value.

Most programmers use the *sizeof* operator to set the *malloc()* function's argument. For example, if you need space to store five *int* values, you construct a *malloc()* statement like this:

```
p = malloc( sizeof(int) * 5);
```

The *sizeof* operator obtains the storage required for a single integer value, multiplied by five, to allocate a 5-integer storage buffer.

Further, programmers with social skills typecast the *malloc()* function to return the data type allocated. Improving upon the preceding example:

```
p = (int *)malloc( sizeof(int) * 5);
```

The *int* typecast includes the * (pointer) operator, to ensure that the memory chunk allocated is of the integer pointer data type.

Finally, you must include the `stdlib.h` header file in your source code to keep the compiler pleased with the *malloc()* function. Listing 20-1 shows an example.

LISTING 20-1: Give Me Space

```
#include <stdio.h>
#include <stdlib.h>

int main()
{
    int *age;

    /* allocate memory */
    age = (int *)malloc(sizeof(int)*1);
    if(age == NULL)
    {
        puts("Unable to allocate memory");
        exit(1);
    }

    /* use the memory */
    printf("How old are you? ");
    scanf("%d",age);
    printf("You are %d years old.\n",*age);

    return(0);
}
```

The first thing to notice about Listing 20-1 is that the only variable declared is *int* pointer, `age`. This pointer isn't assigned the address of another variable, but rather is used to hold the address of memory allocated later in the code.

Line 9 uses *malloc()* to set aside storage for one integer. To ensure that the proper amount of storage is allocated, the *sizeof* operator is used. To allocate space for one integer, the value `1` is multiplied by the result of the `sizeof(int)` operation. The address returned is saved in the `age` pointer.

Line 10 tests to ensure that *malloc()* was successful. If not, the value returned is NULL (a constant defined in `stdlib.h`), and the program outputs an error message (Line 12) and quits (Line 13).

The *scanf()* function at Line 18 doesn't use the `&` prefix for its second argument because the `age` variable is a memory address — a pointer.

Finally, peeker notation is used in Line 19 to output the value input.

Exercise 20-1: Fire up a new program using the source code from Listing 20-1. Build and run.

Exercise 20-2: Using Listing 20-1 as your inspiration, write a program that asks for the current temperature outside as a floating-point value. Use *malloc()* to create storage for the value input. Have the program ask whether the input is Celsius or Fahrenheit. Output the resulting temperature in Kelvin. Here are the formulae:

```
kelvin = celsius + 273.15;

kelvin = (fahrenheit + 459.67) * (5.0/9.0);
```

Exercise 20-3: Write a program that allocates space for three *int* values — an array. You need to use only one *malloc()* function to accomplish this task. Assign 100, 200, and 300 to each *int*, and then display all three values. Use pointer notation throughout the code.

Creating string storage

The *malloc()* function is commonly used to create a text input buffer. This technique avoids declaring and sizing an empty array. For example, the notation

```
char input[64];
```

can be replaced by this statement:

```
char *input;
```

The size of the buffer is set in the code by using the *malloc()* function. In Listing 20-2, *malloc()* at Line 8 declares a *char* array — a storage buffer — for about 1,024 bytes. Okay, it's a kilobyte (KB). I remember when they were a big deal.

LISTING 20-2: Allocating an Input Buffer

```
#include <stdio.h>
#include <stdlib.h>

int main()
{
    char *input;

    /* allocate memory */
    input = (char *)malloc(sizeof(char)*1024);
    if(input==NULL)
    {
        puts("Unable to allocate buffer! Oh no!");
        exit(1);
    }

    /* use the memory */
    puts("Type something long and boring:");
    fgets(input,1024,stdin);
    puts("You wrote:");
    printf("\"%s\"\n",input);

    return(0);
}
```

Lines 9 through 14 in Listing 20-2 allocate a 1K storage buffer. The rest of the code from Lines 17 through 20 accepts input and then displays the output.

Exercise 20-4: Whip up a new program using the source code from Listing 20-2.

Exercise 20-5: Modify the source code from Listing 20-2. Use the *malloc()* function to create a second *char* buffer. After text is read by the *fgets()* function, copy text from the first buffer (`input` in Listing 20-2) into the second buffer — all the text except for the newline character, `\n`, at the end of input. Output the result.

Using the calloc() function

The memory allocated by the *malloc()* function is uninitialized. Whatever garbage values linger in memory remain there when the buffer is created. For this reason you must always initialize the allocated buffer before you use it. An example of how things might go wrong is shown in Listing 20-3.

LISTING 20-3: Show Me the Garbage

```
#include <stdio.h>
#include <stdlib.h>

int main()
{
    unsigned char *junk;
    int x;

    /* allocate memory */
    junk = malloc(64);
    if( junk==NULL )
    {
        puts("Unable to allocate memory");
        exit(1);
    }

    /* output the buffer */
    for(x=0;x<64;x++)
    {
        printf("%02X ",*(junk+x));
        if( (x+1) % 8 == 0 )
            putchar('\n');
    }

    return(0);
}
```

Exercise 20-6: Copy the source code from Listing 20-3 in your editor. Save, build, and run.

The output from Exercise 20-6 shows the 64 bytes allocated as 2-digit hexadecimal values. The values are unpredictable. Memory could contain all zeros or any other values. The point is that the information is raw and uninitialized, and shouldn't be trusted.

When you must ensure that memory is allocated and initialized, use the *calloc()* function instead of *malloc()*. Its format requires two arguments:

```
p = calloc(size,type);
```

The first argument, size, is the quantity of memory requested. The second argument, type, is the size of each item requested. For example:

```
p = calloc(64,sizeof(char));
```

The preceding statement allocates a 64-character memory buffer, all values initialized to \0, the null character.

The following statement sets aside storage for 16 integers, all initialized to zero:

```
v = calloc(16,sizeof(int));
```

Like the *malloc()* function, the *calloc()* function is prototyped in the stdlib.h header file, which must be included in your source code file, lest the compiler get all huffy.

Exercise 20-7: Fix the source code from Listing 20-3 so that the *calloc()* function is used to allocate 64 *char* values. Confirm that the output is all zeros.

Getting more memory

The *malloc()* function has a companion function that should be called *oops()*. Instead, computer scientists determined that it be called *realloc()*. As you might guess, this function's purpose is to reallocate memory, changing the buffer size as needed. Listing 20-4 shows how it works.

LISTING 20-4: Giving Back a Few Bytes

```
#include <stdio.h>
#include <stdlib.h>
#include <string.h>

int main()
{
    char *input;
    int len;

    /* allocate storage */
    input = (char *)malloc(sizeof(char)*1024);
    if(input==NULL)
    {
        puts("Unable to allocate buffer! Oh no!");
        exit(1);
```

(continued)

LISTING 20-4: *(continued)*

```
    }
    /* gather input */
    puts("Type something long and boring:");
    fgets(input,1023,stdin);

    /* resize the buffer */
    len = strlen(input);
    input = realloc(input,sizeof(char)*(len+1));
    if( input==NULL )
    {
        puts("Unable to reallocate buffer!");
        exit(1);
    }
    puts("Memory reallocated.");

    /* output results */
    puts("You wrote:");
    printf("%s",input);

    return(0);
}
```

The source code in Listing 20-4 is based on Listing 20-2, with code added to accommodate the *realloc()* function at Line 24. Here's the format:

```
p = realloc(buffer,size);
```

`buffer` is an existing, allocated storage area, created by the *malloc()* or similar function. `size` is the new buffer size, calculated the same as the *malloc()* function's `size` argument. Upon success, *realloc()* returns a pointer to `buffer`; otherwise, NULL is returned. And, yes, you can use the same pointer variable as both `p` and `buffer`, as shown with variable `input` at Line 24 in Listing 20-4.

As with *malloc()*, the *realloc()* function requires the `stdlib.h` header, shown in Listing 20-4 at Line 2.

The `string.h` header is called in at Line 3 to satisfy the use of the *strlen()* function at Line 23. The input string's length is gathered and saved in the `len` variable.

At Line 24, the *realloc()* function resizes the `input` buffer to a new value. The new value is based on the input string's length plus 1, to account for the `\0` character. The point of this code is to resize the input buffer to match the string's exact length, which is the value of `len` plus one for the null character.

If the *realloc()* function is successful, it resizes the `input` buffer. If it's unsuccessful, a NULL is returned, which is tested for at Line 25.

Exercise 20-8: Type the source code from Listing 20-4 into your editor. Build and run.

Freeing memory

Because C is a mid-level language, a lot of the memory management chore falls upon you, the programmer. When the code is done using allocated memory, it should be freed. This step allows the memory to be used again, and it prevents allocated memory chunks from piling up like empty Amazon boxes in your garage.

The *free()* function is demonstrated in Listing 20-5.

LISTING 20-5: If You Love Your Memory, Set It Free

```
#include <stdio.h>
#include <stdlib.h>

int main()
{
    int *age;

    /* allocate memory */
    age = (int *)malloc(sizeof(int)*1);
    if(age==NULL)
    {
        puts("Out of Memory or something!");
        exit(1);
    }

    /* use memory */
    printf("How old are you in years? ");
    scanf("%d",age);
    *age *= 365;
    printf("You're over %d days old!\n",*age);

    /* free memory */
    free(age);

    return(0);
}
```

The code shown in Listing 20-4 doesn't contain any surprises; most of it should be familiar if you've worked through this chapter straight from the beginning. The only new item is at Line 23, the *free()* function.

The *free()* function releases the allocated memory, making it available for *malloc()* or something else to use. Its argument is a pointer, the address of memory to free. Like *malloc()*, *calloc()*, and *realloc()*, the *free()* function is prototyped in the `stdlib.h` header file.

Exercise 20-9: Type the source code from Listing 20-4 into a new project. Build and run.

You may have noticed that no earlier listings in this chapter use the *free()* function to release allocated memory. The reason is that memory was allocated in the *main()* function, and the operating system releases all the program's memory when the program quits. Even in Listing 20-5, the *free()* function is redundant.

Where the *free()* function is vital is in functions or elsewhere that temporary memory is allocated. When the program is done using the memory, it must be freed or else it accumulates and hogs system resources. An example of using *free()* in this manner is provided later in this chapter, in Listing 20-8.

Lists That Link

At the intersection of Structure Street and Pointer Place, you find a topic known as the *linked list.* It's an array of structures, like a database. The big difference is that each structure is carved out of memory one at a time, like hewing blocks of marble to build an elaborate temple. It's a marvelously nerdy topic and an excellent demonstration of how pointers can be useful. Linked lists put the *malloc()* function to the test.

Allocating space for a structure

As a C language data storage thingy, a structure can be allocated storage just like any other C language data type. The *malloc()* function uses the structure's definition to allocate memory. The address returned is assigned to a structure pointer variable. The operation works as you'd expect — except for one thing: Instead of a period used to reference structure members, allocated structures use the structure pointer operator, which looks like this: ->.

As an example, structure variable date has integer member day. To assign a value to this member, use the following statement:

```
date.day = 14;
```

When structure variable date is a pointer, its day member is accessed like this:

```
date->day = 14;
```

Weird. You might be wondering why the expression isn't *date.day — which is similar to how the expression looked back in the ancient days of the C language:

```
(*date).day = 14;
```

The parentheses are required in order to bind the * pointer operator to date, the structure pointer variable name; otherwise, the . (member) operator takes precedence. But, for some reason, primitive C programmers detested this format, so they went with -> instead.

Listing 20-6 demonstrates how the *malloc()* function allocates storage for a structure. The stk structure is defined at Line 7. Pointer variable invest of the stk structure type is declared at Line 12. In Line 15, *malloc()* allocates storage for one stk structure. The size of the structure is determined by using the *sizeof* operator on the structure definition, not the structure variable name.

LISTING 20-6: Creating a Structured Portfolio

```
#include <stdio.h>
#include <stdlib.h>
#include <string.h>

int main()
{
    struct stk {
        char symbol[5];
        int quantity;
        float price;
    };
    struct stk *invest;

    /* allocate structure */
    invest=(struct stk *)malloc(sizeof(struct stk)*1);
    if(invest==NULL)
```

(continued)

LISTING 20-6: *(continued)*

```
    {
        puts("Some kind of malloc() error");
        exit(1);
    }

    /* assign structure data */
    strcpy(invest->symbol,"GOOG");
    invest->quantity=26;
    invest->price=1373.19;

    /* output database */
    puts("Investment Portfolio");
    printf("Symbol\tShares\tPrice\tValue\n");
    printf("%-6s\t%5d\t%.2f\t%.2f\n",
            invest->symbol,
                invest->quantity,
                invest->price,
                invest->quantity*invest->price);

    return(0);
}
```

The invest pointer references the new structure carved out of memory. Lines 23 through 25 fill the structure with data. Then Lines 28 through 34 output the data. See how the -> operator is used to reference the structure's members?

Exercise 20-10: Create a new program by using the source code from Listing 20-6. Build and run.

TECHNICAL STUFF

The -> operator is used only when the structure is an allocated pointer. For a regular structure with a pointer member, the dot operator is still used. If the structure is a pointer with a pointer member, the -> operator is used.

Creating a linked list

If you wanted to add a second structure to the source code in Listing 20-5, you'd probably create another structure pointer, something like this:

```
struct stk *invest2;
```

You'd probably rename the original invest pointer to invest1 to keep things clear. Then you'd say, "You know, this smells like the start of an array," so you create an array of structure pointers, `struct stk invest[]`. Yes, sir, all of that works.

What works better, however, is to create a linked list, a series of structures that contain pointers referencing other structures in the list. To make this change, one new member is added to the `stk` structure: a pointer to the `stk` structure.

Yes, this concept is difficult to describe. So, rather than spin words, look at Listing 20-7.

LISTING 20-7: A Primitive Linked-List Example

```
#include <stdio.h>
#include <stdlib.h>
#include <string.h>

int main()
{
    struct stk {
        char symbol[5];
        int quantity;
        float price;
        struct stk *next;
    };
    struct stk *first;
    struct stk *current;

    /* allocate structre */
    first=(struct stk *)malloc(sizeof(struct stk)*1);
    if(first==NULL)
    {
        puts("Some kind of malloc() error");
        exit(1);
    }
    /* set the base */
    current=first;

    /* assign structure data */
    strcpy(current->symbol,"GOOG");
    current->quantity=26;
    current->price=1373.19;

    /* allocate the next structure */
    current->next=\
```

(continued)

LISTING 20-7: (continued)

```
        (struct stk *)malloc(sizeof(struct stk)*1);
      if(current->next==NULL)
    {
       puts("Another malloc() error");
       exit(1);
    }
    /* reference the new structure */
    current=current->next;
    strcpy(current->symbol,"MSFT");
    current->quantity=67;
    current->price=183.16;
    /* cap the end of the list */
    current->next=NULL;

    /* outupt database */
    puts("Investment Portfolio");
    printf("Symbol\tShares\tPrice\tValue\n");
    current=first;
    while( current )
    {
        printf("%-6s\t%5d\t%.2f\t%.2f\n",
                current->symbol,
                current->quantity,
                current->price,
                current->quantity*current->price);
        current=current->next;
    }

    return(0);
}
```

The source code shown in Listing 20-7 is long, but it's based on Listing 20-6. I just created a second structure, linked to the first one. So don't let the source code's length intimidate you.

Key to the linked list are three required structure pointer variables: `first`, `current`, and `next`.

Lines 13 and 14 declare two of the traditional structure pointers, `first` and `current`. The third pointer variable, `next`, is incorporated into the linked list structure at Line 11. Before moving on, a couple of warnings:

WARNING

Do not use *typedef* to define a new structure variable when creating a linked list. I'm not using *typedef* in Listing 20-7, so it's not an issue with the code, but many C programmers use *typedef* with structures. Be careful!

WARNING

The third traditional linked-list variable is named `new`. I use `next` instead because `new` is a reserved word in C++. Though you can use `new` in your C code, it's best not to confuse the two. Therefore, I've rewritten the code to use the `next` variable name, which is also part of the linked-list structure shown in Listing 20-7.

At Line 17, the `first` pointer is allocated. This pointer forms the base address of the linked list, and it must never be altered or else the entire list is lost. Therefore, at Line 24, it's assigned to the `current` pointer, which is used to assign values in the current list structure.

At Line 32, the `next` member of the current structure is allocated to a new `stk` structure in memory. (The line is split so that it doesn't mess up this book's layout.) Upon success, at Line 40 the `current` structure pointer is assigned to `current->next`, and this new structure's data is filled.

The linked list is capped at Line 45. The NULL pointer is assigned to the last structure's `next` member.

Line 50 resets the list to the base, assigning pointer `current` to `first`. A *while* loop outputs the list as long as the `current` pointer isn't NULL. This pointer is updated at Line 58.

Exercise 20-11: Type the source code from Listing 20-7 into your editor, or just copy over the source code from Exercise 20-10 and modify it. Even though it's long, type it in because you need to edit it again later (if you're not used to that by now). Build and run.

Figure 20-1 illustrates the concept of the linked list based on what Listing 20-7 does.

FIGURE 20-1: A linked list in memory.

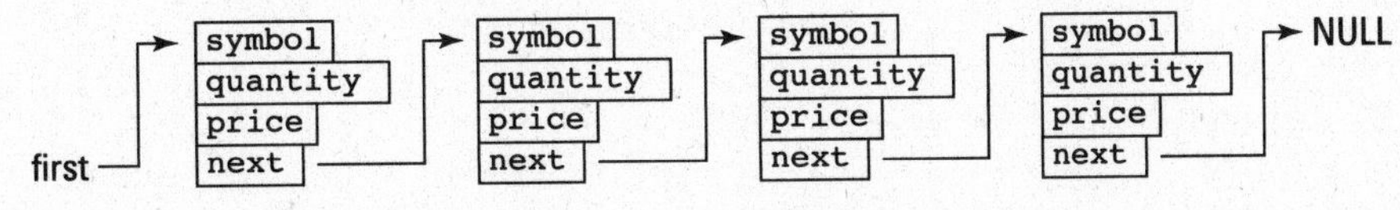

Unlike arrays, structures in a linked list aren't numbered. Instead, each structure is linked to the next one in the list. Providing that you know the address of the first structure, you can work through the list until the end, which is marked by a NULL pointer.

Listing 20-6 could use some cleaning. Functions scream at me, so I'll heed their cries with an improved version of the code, shown in Listing 20-8.

LISTING 20-8: A Better Linked-List Example

```
#include <stdio.h>
#include <stdlib.h>
#include <string.h>

#define ITEMS 5

struct stk {
    char symbol[5];
    int quantity;
    float price;
    struct stk *next;
};

struct stk *make_structure(void);
void fill_structure(struct stk *a,int c);
void show_structure(struct stk *a);

int main()
{
    struct stk *first;
    struct stk *current;
    int x;

    /* create the ITEMS-sized linked list */
    for(x=0;x<ITEMS;x++)
    {
        if(x==0)
        {
            first=make_structure();
            current=first;
        }
        else
        {
            current->next=make_structure();
            current=current->next;
        }
        fill_structure(current,x+1);
    }
    /* cap the list */
    current->next=NULL;
```

```
    /* output the database */
    puts("Investment Portfolio");
    printf("Symbol\tShares\tPrice\tValue\n");
    current = first;
    while(current)
    {
        show_structure(current);
        current=current->next;
    }

    return(0);
}

/* allocate a new structure */
struct stk *make_structure(void)
{
    struct stk *a;

    a=(struct stk *)malloc(sizeof(struct stk)*1);
    if(a==NULL)
    {
        puts("Some kind of malloc() error");
        exit(1);
    }

    return(a);
}

/* fill the structure with data */
void fill_structure(struct stk *a,int c)
{
    printf("Item #%d/%d:\n",c,ITEMS);
    printf("Stock Symbol: ");
    scanf("%s",a->symbol);
    printf("Number of shares: ");
    scanf("%d",&a->quantity);
    printf("Share price: ");
    scanf("%f",&a->price);
}

/* output the structure */
void show_structure(struct stk *a)
```

(continued)

LISTING 20-8: (continued)

```
{
    printf("%-6s\t%5d\t%.2f\t%.2f\n",\
            a->symbol,
            a->quantity,
            a->price,
            a->quantity*a->price);
}
```

Most linked lists are created as shown in Listing 20-8. The key is to use three structure variables, shown at Lines 11, 20, and 21:

- next is a structure member that references the next structure in the list. It's allocated with a new structure as the program runs. This pointer is capped with the NULL constant at the end of list.
- first always contains the address of the first structure in the list. Always.
- current contains the address of the structure being worked on, filled with data, or output.

Line 7 declares the stk structure as global so that it can be accessed from the various functions.

The *for* loop between Lines 25 and 38 creates new structures, linking them together. The initial structure is special, so its address is saved in Line 29. Otherwise, a new structure is allocated, thanks to the *make_structure()* function.

In Line 34, a new structure is allocated using the next pointer member of the current structure. The current pointer is updated to reference the freshly allocated structure at Line 35.

At Line 40, the end of the linked list is marked by resetting the current pointer in the last structure to a NULL.

The *while* loop at Line 46 outputs all structures in the linked list. The loop's condition is the value of the current pointer. When the NULL is encountered, the loop stops.

The rest of the code shown in Listing 20-8 consists of functions that are self-explanatory. If the *make_structure()* function concerns you because you believe the value of variable a to be lost, you're correct. Its value *is* lost, but the address returned from the function continues to reference an allocated chunk of memory.

Exercise 20-12: Copy the code from Listing 20-8 into the editor. Build and run.

REMEMBER

Take note of the *scanf()* statements in the *fill_structure()* function. Remember that the -> is the "peeker" notation for a pointer. To get the address, you must prefix the variable with an & in the *scanf()* function.

Editing a linked list

Because a linked list is chained together by referencing memory locations, editing is done by modifying those memory locations. For example, in Figure 20-2, if you want to remove the third item from the list, you dodge around it by linking the second item to the fourth item. The third item is effectively removed (and lost) by this operation.

Likewise, you can insert an item into the list by editing the next pointer from the previous structure, as illustrated in Figure 20-3.

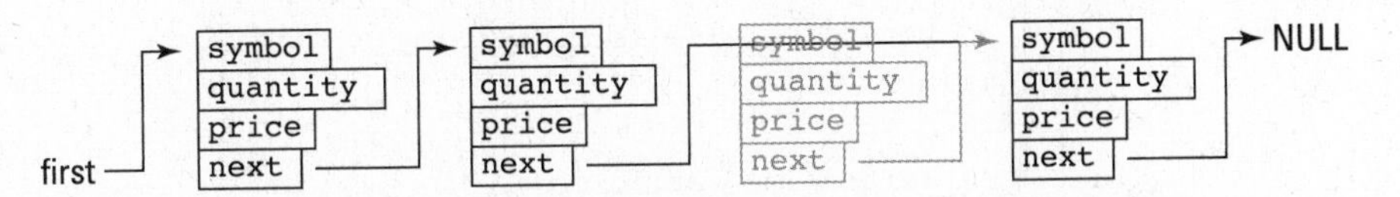

FIGURE 20-2: Removing an item from a linked list.

The best way to alter items in a linked list is to have an interactive program that lets you view, add, insert, delete, and edit the various structures. Such a program would be quite long and complex, which is why it's shown in Listing 20-9.

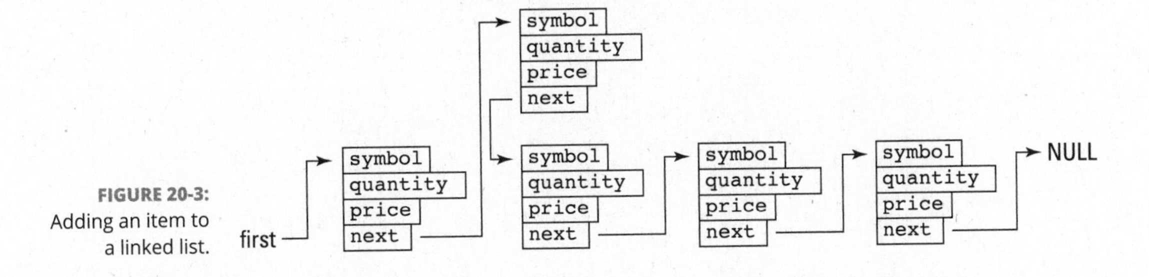

FIGURE 20-3: Adding an item to a linked list.

LISTING 20-9: An Interactive Linked-List Program

```
/* An interactive linked-list program */
/* Dan Gookin, C Programming For Dummies */
#include <stdio.h>
#include <stdlib.h>
#include <ctype.h>

struct lili {
    int value;
    struct lili *next;
};
struct lili *first;
struct lili *current;

int menu(void);
void add(void);
void show(void);
void delrec(void);
struct lili *create(void);

/* the main function works with input only
      everything else is handled by a function */
int main()
{
    /* initialize variables */
    int choice='\0';
    first=NULL;

    while(choice!='Q')
    {
        choice=menu();
        switch (choice)
        {
            case 'S':
                show();
                break;
            case 'A':
                add();
                break;
            case 'R':
                delrec();
                break;
            case 'Q':
                break;
```

```
        }
    }

    return(0);
}

/* output the main menu and collect input */
int menu(void)
{
    int ch;

    printf("S)how, A)dd, R)emove, Q)uit: ");
    ch=getchar();
    /* delete excess input */
    while(getchar()!='\n')
        ;
    return(toupper(ch));
}

/* add an item to the end of the linked list */
void add(void)
{
    /* special case for the first item */
    if(first==NULL)
    {
        first=create();
        current=first;
    }
    /* otherwise, find the last item */
    else
    {
        current=first;
        /* find the NULL */
        while(current->next)
            current=current->next;
        current->next=create();
        current=current->next;
    }
    printf("Type a value: ");
    scanf("%d",&current->value);
    current->next=NULL;
    /* delete excess input */
    while(getchar()!='\n')
        ;
}
```

(continued)

LISTING 20-9: ***(continued)***

```
/* output all structures in the linked list */
void show(void)
{
    int count=1;

    /* is the list empty? */
    if(first==NULL)
    {
        puts("Nothing to show");
        return;
    }
    puts("Showing all records:");
    current=first;
    /* loop until current==NULL */
    while(current)
    {
        printf("Record %d: %d\n",
                count,current->value);
        current=current->next;
        count++;
    }
}

/* delete a record from the list */
void delrec(void)
{
    /* the previous record must be saved */
    struct lili *previous;
    int r,c;

    /* is the list empty? */
    if(first==NULL)
    {
        puts("No records to remove");
        return;
    }

    puts("Choose a record to remove:");
    /* output the list */
    show();
    printf("Record: ");
    scanf("%d",&r);
    /* delete excess input */
```

```
    while(getchar()!='\n')
        ;
    c=1;
    /* reset the list */
    current=first;
    /* initialize the previous pointer */
    previous=NULL;
    while(c!=r)
    {
        previous=current;
        current=current->next;
        c++;
        /* bail on the end of the list */
        if(current==NULL)
        {
            puts("Record not found");
            return;
        }
    }
    /* 'current' holds the record to delete */
    /* is it the first record? */
    if(previous==NULL)
        first=current->next;
    /* otherwise, reset the next reference */
    else
    previous->next=current->next;
    printf("Record %d removed.\n",r);
    /* release memory of the current record */
    free(current);
}

/* build an empty structure & return its address */
struct lili *create(void)
{
    struct lili *a;

    a=(struct lili *)malloc(sizeof(struct lili)*1);
    if(a==NULL)
    {
        puts("Some kind of malloc() error");
        exit(1);
    }

    return(a);
}
```

Exercise 20-13: If you have the time, type the source code from Listing 20-9 into your editor. I could argue that typing it in helps you better understand the code. Build and run a few times to get the hang of it.

Saving a linked list

Linked lists exist only in memory. Though you can save all records from a linked list to a file, the `next` pointer member in each structure should be discarded when the information is read from the file. The reason is because the linked list may not lay out in the same chunk of memory.

See Chapter 22 for details on working with files — specifically, the topic of random file access.

» Understanding the Unix epoch

» Retrieving the current time

» Displaying the date and time

» Pausing program execution

Chapter 21
It's About Time

It's time to program! Or, to put it another way, it's time to program time. The C library is bursting with various time-oriented functions, allowing you to not only report the current time but also display dates and times. You can even suspend a program's execution — on purpose — providing you know the proper functions.

What Time Is It?

Does anyone have the time? Seriously, does anyone really know what time it is — or was?

Electronic devices have clocks, but this feature doesn't make them the best time-keepers. In fact, most gizmos today constantly update their internal clocks by using an Internet time server. Otherwise, the clock on your computer, cell phone, or tablet would never be accurate.

When you program the time in C, you're relying upon the device you're using to accurately report the date and time. That brings to light a whole bunch of terms and technology surrounding the subject of time and how it's measured.

Understanding the calendar

Digital devices keep track of time by counting ones and zeros. Humans like to keep track of time by counting seconds, minutes, hours, days, weeks, months, and years. Various schemes have been developed to work between the two systems.

The *Julian calendar* was popular for centuries. Developed by Julius Caesar and programmed in Latin, this calendar worked well for a long time.

Sadly, old Julius didn't account for fractions of a day that accumulated over time. In the year 1500, Pope Gregory developed the *Gregorian calendar*, which fixed Caesar's oversights. This calendar was also programmed in Latin.

Computer scientists developed something called the *Modified Julian Date* (MJD), back in the 1950s. They set the date January 1, 4713 B.C. as Day 0 and numbered each day since. Hours are given fractional parts of the day. Noon on January 1, 2014, was 2456293.5 MJD.

When Unix popped into being, two things were born: the C language and the Unix epoch. At midnight on January 1, 1970, Unix computers started counting the seconds. The *Unix epoch* is measured since that moment as a *time_t* value, which is a *typedef* of a *long unsigned int*. This data type's range makes the Unix epoch calendar valid until January 19, 2038, at 3:14:07 a.m., when the computer suddenly believes that the date is December 13, 1901, all over again. And that was a Friday!

Most Unix computers have addressed the 2038 problem, so unlike the Y2K crisis, nothing bad happens after January 19, 2038. The Unix epoch, however, is still used in time programming C functions.

Working with time in C

Time functions and related matters in the C language are held in the `time.h` header file. In this file, you find goodies to sate your program's timely needs. These items include the *time_t* data type and `tm` structure:

time_t	The *time_t* data type holds the value of the Unix epoch, or the number of seconds that have passed since January 1, 1970. On most systems, *time_t* is a *typedef* of a *long signed int*; use the `%ld` or `%lu` placeholder to output this value.
`struct tm`	The `tm` structure holds definitions for storing various parts of a timestamp. It's filled by the *localtime()* function. See the later section "Slicing through the time string."

Also defined in the `time.h` header file are these chronological functions:

time()	The *time()* function uses the address of the *time_t* variable as its argument. This function fills the *time_t* variable with the current Unix epoch time and it also returns a *time_t* value.
ctime()	The *ctime()* function consumes a *time_t* variable and converts it into a displayable date-time string.
localtime()	This function fills a `tm` structure with information based on the time value stored in a *time_t* variable. The *localtime()* function returns the address of the `tm` structure, so it gets all messy with structures and pointers and that -> operator.
difftime()	The *difftime()* function compares the values between two *time_t* values and returns a *float* value as the difference in seconds.

C libraries often come with other time-related functions, such as *sleep()* and *clock()*. These functions may not be available on all systems — which shouldn't be a problem, because you can always code your own time functions.

Time to Program

I could imagine that the same Programming Lords who invented pointers and linked lists could really mess up your brain with time programming. Happily, they didn't. Though a knowledge of pointers and structures helps you learn the ropes, time programming in C is straightforward.

Checking the clock

The computer, or whichever device you're programming, keeps a time value somewhere deep in its digital bosom. To access this value, you need to code a program that uses the *time()* function. It stuffs the current clock tick value into a *time_t* variable, as shown in Listing 21-1.

LISTING 21-1: Oh, So This Is the Time?

```
#include <stdio.h>
#include <time.h>

int main()
{
    time_t tictoc;

    time(&tictoc);
    printf("The time is now %ld\n",tictoc);
    return(0);
}
```

Line 2 in Listing 21-2 brings in the `time.h` header file, which is required for various time functions in C.

Line 6 declares the *time_t* variable `tictoc`. The *time()* function at Line 8 requires the address of a *time_t* variable as its argument. The time value is placed directly into variable `tictoc` (in its memory location).

Finally, in Line 9, the resulting value — the Unix epoch — is output by using the `%ld` conversion character, *long int*.

Exercise 21-1: Type the source code from Listing 21-1 into your editor. Build and run.

Exercise 21-2: Edit your solution from Exercise 21-1, replacing Line 8 with

```
tictoc=time(NULL);
```

The *time()* function requires a memory location as an argument, but it also returns a *time_t* value. You can use either the format just presented or the format shown at Line 8 in Listing 21-1, depending on which weirdo symbol, the & or NULL, frightens you the most.

TECHNICAL STUFF

Back in the 1970s and '80s, BASIC programmers would write *for* loops in their code to pause program execution. I recall that my trusty old TRS-80 required a loop that counted from 1 to 100 to delay execution by one second. Today's systems are far faster, and such loops can no longer be relied upon to accurately delay program execution.

HOW *TIME()* PLAYS INTO RANDOM NUMBERS

The best way to generate random numbers in C is to seed the randomizer. In Chapter 11, I describe how that process works by using the *time()* function. Here's the format:

```
srand((unsigned)time(NULL));
```

When called with a NULL value (a pointer), the *time()* function returns the current time of day in the Unix epoch format. To ensure that the value returned — the Unix epoch — isn't negative, it's typecast to an *unsigned* type. This step may not be required, because the *time_t* value may be *unsigned* on certain systems. Still, the typecast ensures that the value is always positive.

Viewing a timestamp

Displaying the current date-and-time as an integer won't make your users happy. In fact, I don't even know any Unix geeks who can look at a Unix epoch number and determine which date it is. Therefore, some conversion needs to take place. The C library function required to fulfill that duty is *ctime()*, the time conversion function, shown in Listing 21-2.

LISTING 21-2: Oh, So That's the Time!

```
#include <stdio.h>
#include <time.h>

int main()
{
    time_t tictoc;

    time(&tictoc);
    printf("The time is now %s",ctime(&tictoc));
    return(0);
}
```

The *ctime()* function eats the address of a *time_t* variable. The function returns a pointer to a timestamp string, which is the address of a *char* array elsewhere in memory. The string includes a terminating newline.

Exercise 21-3: Create a new program using the source code from Listing 21-2. Don't forget the `%s` conversion character in the *printf()* statement. Build and run. The output looks something like this:

```
The time is now Mon May 18 13:54:31 2020
```

The latter part of the string (starting with `Mon`) is returned from the *ctime()* function.

Exercise 21-4: Write code that passes the *time_t* value 946684800 to the *ctime()* function. Output the string that's returned.

Slicing through the time string

Don't fret about the string that *ctime()* returns, or about your code having to slice through it to obtain specific date or time tidbits. That's because the *localtime()* function can be used with a *time_t* value to squeeze out individual bits and pieces of the current time.

The *localtime()* function extracts time-and-date information based on a *time_t* value. It fills the relevant parts into a `tm` structure, which looks something like this:

```
struct tm {
    int tm_sec;    /* seconds after the minute [0-60] */
    int tm_min;    /* minutes after the hour [0-59] */
    int tm_hour;   /* hours since midnight [0-23] */
    int tm_mday;   /* day of the month [1-31] */
    int tm_mon;    /* months since January [0-11] */
    int tm_year;   /* years since 1900 */
    int tm_wday;   /* days since Sunday [0-6] */
    int tm_yday;   /* days since January 1 [0-365] */
    int tm_isdst;  /* Daylight Saving Time flag */
};
```

Listing 21-3 shows how to use the `tm` structure to output a recognizable date format.

LISTING 21-3: What's Today?

```
#include <stdio.h>
#include <time.h>

int main()
{
    time_t tictoc;
    struct tm *today;

    time(&tictoc);
    today = localtime(&tictoc);
    printf("Today is %d/%d/%d\n",
        today->tm_mon,
        today->tm_mday,
        today->tm_year);
    return(0);
}
```

The current time in Unix epoch ticks is gathered by the *time()* function in Line 9 in Listing 21-3. The value is stored in the `tictoc` variable.

At Line 10, the *localtime()* function returns the `tm` structure pointer `today`. The structure's elements are output by *printf()* across a few lines. Structure pointer notation is used to access the structure's elements (Lines 12, 13, and 14) because the structure is a pointer (memory address).

Exercise 21-5: Type the source code from Listing 21-3 into your editor. Build and run the program to see the current date.

Because Trajan is no longer the Roman emperor and it isn't last month, you must make some adjustments to the code. Refer to the definition for the structure earlier in this section and you'll understand the math necessary to output the proper month-and-year values.

Exercise 21-6: Fix your solution to Exercise 21-5 so that the current year and current month are output.

Exercise 21-7: Write code that outputs the current time in the format `hour:minute:second`. Ensure that the minute-and-second output is two digits wide with a leading zero.

Exercise 21-8: Fix your solution from Exercise 21-7 so that the output is in 12-hour format with an A.M. or P.M. suffix based on the time of day.

Exercise 21-9: Write code that outputs the full name for the current day of the week.

Snoozing

Most programmers want their code to run fast. Occasionally, you want your code to slow down, to take a measured pause, or to build . . . suspense! In these instances, you can rely upon C's time functions to cause a wee bit of delay.

The *sleep()* function is a common, though nonstandard, time function. It's prototyped in the `unistd.h` header file. The function's argument is the number of seconds to pause program execution.

The *difftime()* function, shown in Listing 21-4, is prototyped in the `time.h` header. Use it to calculate the difference between the two *time_t* values, `now` and `then`, in the code. The function returns a floating-point value indicating the number of seconds elapsed.

LISTING 21-4: Wait a Sec!

```
#include <stdio.h>
#include <time.h>

int main()
{
    time_t now,then;
    float delay=0.0;

    time(&then);
    puts("Start");
    while(delay < 1)
    {
        time(&now);
        delay = difftime(now,then);
        printf("%f\r",delay);
    }
    puts("\nStop");
    return(0);
}
```

Exercise 21-10: Type the source code from Listing 21-4 into your editor. Build the program. Run.

5

And the Rest of It

IN THIS PART . . .

Read and write information to files

Desperately save a linked list

Use C to perform file management

Create larger projects

Rid your code of bugs

» Reading and writing text to a file

» Creating binary files

» Using the *fread()* and *fwrite()* functions

» Reading and writing records

» Building a linked-list database

Chapter 22

Permanent Storage Functions

C programs work innately with storage in memory. Variables are created, values are set, locations are mapped. It's all pretty much automatic, but the information that's created is lost after the program runs.

For the long term, programs need to access permanent storage, writing and reading information to and from files. C comes with a host of interactive functions that let you create, read, write, and manipulate files. These are the *permanent storage functions.*

Sequential File Access

The simplest way that information can be written to or read from a file is *sequentially,* one byte after the other. The file contains one long stream of data. This information is accessed from start to finish, like watching a movie on broadcast TV.

Understanding C file access

File access in C is another form of I/O. Rather than use standard input (the keyboard) or standard output (the display), the input or output acts upon a file. Sounds simple. Happily, it is.

A file is opened by using the *fopen()* function:

```
handle = fopen(filename,mode);
```

The *fopen()* function requires two arguments, both strings. The first is a `filename`; the second is a `mode`. The choices for `mode` are listed in Table 22-1. The *fopen()* function returns a file `handle`, which is a pointer used to reference the file. This pointer is of the *FILE* data type, which is a structure defined in the `stdio.h` header file.

TABLE 22-1: Access Modes for the *fopen()* Function

Mode	File Open for	Create File?	Notes
"a"	Appending	Yes	It adds to the end of an existing file; a file is created if it doesn't exist.
"a+"	Appending and reading	Yes	Information is added to the end of the file.
"r"	Reading	No	If the file doesn't exist, *fopen()* returns an error.
"r+"	Reading and writing	No	If the file doesn't exist, an error is returned.
"w"	Writing	Yes	The existing file is overwritten if the same name is used.
"w+"	Writing and reading	Yes	The existing file is overwritten.

REMEMBER

The *mode* is a string. Even when only one character is specified, it must be enclosed in double quotes.

After the file is open, you use the `handle` variable to reference the file as you read and write. Many file I/O functions are similar to their standard I/O counterparts, but with an *f* prefix. To write to a file, you can use the *fprintf()*, *fputs()*, *fputchar()*, and similar functions. Reading from a file uses the *fscanf()*, *fgets()*, and so on.

After all the reading and writing, the file is closed by using the *fclose()* function with the file handle as its argument.

All the file I/O functions mentioned in this chapter are prototyped in the `stdio.h` header file.

Writing text to a file

To write text to a file, you abide by these steps:

1. **Open the file.**
2. **Check to confirm whether the file opened successfully.**
3. **Write data to the file.**
4. **Close the file.**

Listing 22-1 performs these steps. The file `hello.txt` is created. It's a text file, with the contents `Look what I made!`.

LISTING 22-1: Write That File

```
#include <stdio.h>
#include <stdlib.h>

int main()
{
    FILE *fh;

    /* open the file */
    fh=fopen("hello.txt","w");
    /* check for errors */
    if(fh==NULL)
    {
        puts("Can't open that file!");
        exit(1);
    }
    /* output text */
    fprintf(fh,"Look what I made!\n");
    /* close the file */
    fclose(fh);
    /* inform the user */
    puts("File written.");
    return(0);
}
```

Line 6 creates the file handle, `fh`. It's a pointer. The pointer stores the return value from the *fopen()* function at Line 9. The function's first argument is the filename `hello.txt`. The second argument is `"w"`, the write mode. If the file exists, it's overwritten.

The *if* statement at Line 11 confirms that the file was properly opened. If it wasn't, the value of fh is NULL and appropriate action is taken.

The *fprintf()* function writes text to the file at Line 17. The format is the same as for *printf()*, but with the file handle as the first argument, followed by the format string and then any variables.

Finally, Line 19 closes the file by using the *fclose()* function. This statement is a required step for any file access programming.

Exercise 22-1: Copy the source code from Listing 22-1 into your editor. Build and run the program.

Use your computer's file browser to locate the file that's created and open it. At the bash prompt, use the command **cat hello.txt** to view the file's contents.

Reading text from a file

To read text from a file, you follow the same rules as for writing text to a file, but with reading the text as the third step. (Refer to the preceding section.) Familiar functions are used to access the file's information, but with the *f* prefix and a file handle argument somewhere in the function's parentheses.

To read text from a file one character at a time, use the *fgetc()* function as shown in Listing 22-2.

LISTING 22-2: Read That File

```
#include <stdio.h>
#include <stdlib.h>

int main()
{
    FILE *fh;
    int ch;

    /* open the file */
    fh=fopen("hello.txt","r");
    /* check for errors */
    if(fh==NULL)
    {
        puts("Can't open that file!");
        exit(1);
    }
```

```
    /* loop until the end-of-file */
    while( !feof(fh) )
    {
        /* read one character */
        ch=fgetc(fh);
        /* end of file? */
        if( ch==EOF )
            break;
        /* output character */
        putchar(ch);
    }
    /* close the file */
    fclose(fh);
    return(0);
}
```

Line 11 in Listing 22-2 opens the file `hello.txt` for reading only. The file must exist; otherwise, an error occurs.

The *while* loop at Line 18 repeats based on the return value of the *feof()* function. This function returns TRUE when the end of the file is encountered. The `!` (not) operator reverses this condition, so the loop repeats "while not true," or until all information has been read from the file.

Line 21 uses the *fgetc()* function to read one character from the file identified by handle `fh`. This character is stored in variable `ch`.

At Line 23, the character read from the file stored in variable `ch` is compared with the `EOF`, or End of File, constant. When TRUE, the file's text has been completely read and the *while* loop breaks. `EOF` is an integer constant, which is why variable `ch` is declared as an *int* value. And it's important to run this test in the loop, because the EOF may be countered here before it's read by the *feof()* function at Line 18.

The character read from the file is output at Line 26.

Exercise 22-2: Create a new program by using the source code shown in Listing 22-2. Build and run.

The program outputs the contents of the file created by Exercise 22-1; otherwise, you see the error message.

Exercise 22-3: Modify your source code from Exercise 22-1 to write a second string to the file. Add the following statement after Line 17:

```
fputs("My C program wrote this file.\n",fh);
```

Unlike the *puts()* statement for standard output, you must specify a newline character for *fputs()* file output. Further, the file handle argument appears as the second argument, which is unusual for a C language file function.

Build and run Exercise 22-3, and then rerun your solution from Exercise 22-2 to view the file's contents.

The two functions *fprintf()* and *fputs()* write text sequentially to the file, one character after the other. The process works just like sending text to standard output, but instead the characters are sent to a file in permanent storage.

The *fgets()* function reads an entire string of text from a file, just as it's been used elsewhere in this book to read from standard input (`stdin`). To make it work, you need an input buffer, the number of characters to read, and the file handle. Listing 22-3 shows an example.

LISTING 22-3: Gulping Strings of Text

```
#include <stdio.h>
#include <stdlib.h>

int main()
{
    FILE *fh;
    char buffer[64];
    char *r;

    fh=fopen("hello.txt","r");
    if(fh==NULL)
    {
        puts("Can't open that file!");
        exit(1);
    }
    while( !feof(fh) )
    {
        /* fgets returns (char *)NULL on
           error or EOF */
        r = fgets(buffer,64,fh);
        if( r==NULL )
            break;
```

```
        printf("%s",buffer);
    }
    fclose(fh);
    return(0);
}
```

The *fgets()* function at Line 20 reads in a line of text (terminated by a newline) or 64 characters, whichever comes first. The function returns a pointer to the string read unless an error occurs, in which case NULL is returned. This value is saved in *char* pointer variable `r`, which is tested at Line 21.

Exercise 22-4: Type the source code from Listing 22-3 into your editor. Build and run.

Appending text to a file

When you're using the *fopen()* function in the `"a"` mode, text is appended to an already existing file. Or when the file doesn't exist, the *fopen()* command creates a new file. At Line 9 in Listing 22-4, the "a" mode opens an existing file, `hello.txt`, for appending. If the file doesn't exist, it's created.

LISTING 22-4: Add More Text

```
#include <stdio.h>
#include <stdlib.h>

int main()
{
    FILE *fh;

    /* "a" == append */
    fh=fopen("hello.txt","a");
    if(fh==NULL)
    {
        puts("Can't open that file!");
        exit(1);
    }
    fprintf(fh,"This text was added later\n");
    puts("Text appended");
    fclose(fh);
    return(0);
}
```

The standard file-writing functions are used to spew text to the open file, as shown in Line 15.

Exercise 22-5: Create a new project by using the source code shown in Listing 22-4. Build and run to append text to the `hello.txt` file. Use the program from Exercise 22-4 to view the file's contents. Then run the program (from Exercise 22-5) again to append the text to the file a second time. View the result.

When you're done, the `hello.txt` file contains the following text:

```
Look what I made!
My C program wrote this file.
This text was added later
This text was added later
```

Writing binary data

The demo programs shown so far in this chapter deal with plain text files. Not every file is text, however. Most files contain binary data, unreadable by humans. You can use the C language to work with these files as well. Start by studying Listing 22-5.

LISTING 22-5: Writing Binary Data

```
#include <stdio.h>
#include <stdlib.h>

int main()
{
    FILE *handle;
    int highscore;

    handle = fopen("scores.dat","w");
    if(!handle)
    {
        puts("File error!");
        exit(1);
    }
    printf("What is your high score? ");
    scanf("%d",&highscore);
    fprintf(handle,"%d",highscore);
    fclose(handle);
    puts("Score saved");
    return(0);
}
```

Exercise 22-6: Type the source code from Listing 22-5 into a new project. Build and run.

Most everything that goes on in Listing 22-5 is familiar to you. The problem? Binary data wasn't written. Instead, the *fprintf()* function at Line 17 writes the *int* value to the file as a text string. To prove it, examine the contents of the `scores.dat` file and you see that the value is stored as plain text.

Exercise 22-7: Replace Line 17 from the code for your solution to Exercise 22-6 with this statement:

```
fwrite(&highscore,sizeof(int),1,handle);
```

Save the change. Build and run. When you try to examine the contents of the `scores.dat` file now, it isn't plain text. That's because binary information was written, thanks to the *fwrite()* function.

Here's the format for *fwrite()*:

```
fwrite(variable_ptr,sizeof(type),count,handle);
```

The *fwrite()* function is concerned with writing chunks of information to a file. Unlike the *fprintf()* and *fputs()* functions, it doesn't always write text.

`variable_ptr` is the address of a variable — a pointer. Most programmers satisfy this requirement by prefixing a variable's name with the `&` operator.

Arrays and strings need not be prefixed with the `&` operator when specified in the *fwrite()* function.

`sizeof(type)` is the variable's storage size based on the type of variable — *int*, *char*, and *float*, for example.

`count` is the number of items to write. If you were writing an array of ten *int* values, you specify 10 as the size.

`handle` is the file handle address, returned from an *fopen()* function.

Exercise 22-8: Using the source code from Listing 22-5 as a start, write a program that saves an array of five high-score values to the file `scores.dat`. (It's okay if the program overwrites the original `scores.dat` file.)

The next section covers how to read binary data from a file.

In earlier versions of the C standard, it was common to add a `b` to the *fopen()* function's mode string when reading or writing binary data. This requirement is no longer the case. A modern compiler ignores the `"b"` in a mode string if used in the *fopen()* function.

Reading binary data

The *fread()* function reports for duty when it comes time to read binary information from a file. Like the *fwrite()* function, *fread()* takes in raw data and stuffs it into a C language variable for further examination. Listing 22-6 provides a demonstration. The `scores.dat` file in the listing is created in the preceding section's Exercise 22-8.

LISTING 22-6: Check Those High Scores

```
#include <stdio.h>
#include <stdlib.h>

int main()
{
    FILE *handle;
    int highscore[5];
    int x;

    handle = fopen("scores.dat","r");
    if(!handle)
    {
        puts("File error!");
        exit(1);
    }
    fread(highscore,sizeof(int),5,handle);
    fclose(handle);
    for(x=0;x<5;x++)
        printf("High score #%d: %d\n",x+1,highscore[x]);
    return(0);
}
```

Thanks to the flexibility of the *fread()* function, and its ability to devour multiple values at a time, Line 16 in Listing 22-6 gobbles up all five *int* values that were previously saved in the `scores.dat` file. The *fread()* function works just like *fwrite()* and has the same arguments in the same order, but binary data is read from the file.

In Line 16, the base address of the `highscore` array is passed to *fread()* as the first argument. Next comes the size of each element to be read, the size of an *int* variable. Then comes the immediate value `5`, effectively ordering *fread()* to scan in five values. The final argument is the file handle variable, confusingly named `handle`.

Exercise 22-9: Type the source code from Listing 22-6 into your editor. Build and run to see the five *int* values that were previously saved to the `scores.dat` file.

Because *fread()* can read any file's data, you can use it to create a file-dumper type of program, as shown in Listing 22-7.

LISTING 22-7: A File Dumper

```
#include <stdio.h>
#include <stdlib.h>

int main()
{
    char filename[255];
    FILE *dumpme;
    int x,c;

    printf("File to dump: ");
    scanf("%s",filename);
    dumpme=fopen(filename,"r");
    if(!dumpme)
    {
        printf("Unable to open '%s'\n",filename);
        exit(1);
    }
    x=0;
    while( (c=fgetc(dumpme)) != EOF)
    {
        printf("%02X ",c);
        x++;
        if(!(x%16))
            putchar('\n');
    }
    putchar('\n');
    fclose(dumpme);
    return(0);
}
```

The source code from Listing 22-7 displays each byte in a file. It uses a 2-digit hexadecimal format, `%02X`, to represent each byte in the *printf()* statement at Line 24.

The *fgetc()* function reads the file one byte at a time in Line 21. To prevent the code from reading beyond the end of the file, this byte is compared with the EOF, or end-of-file, marker at Line 22.

The *if* decision at Line 26 uses the modulus operator to determine when 16 bytes have been displayed. When true, a newline is output, keeping the display neat and tidy.

Unlike other programs presented in this chapter, Listing 22-7 prompts the user for a filename at Line 10. Therefore, a good possibility exists that you'll see the error message displayed when an improper filename is typed.

Exercise 22-10: Type the source code from Listing 22-7 into your editor. Build the project. Run it using the `scores.dat` file you created earlier in this chapter, or use the file's own source code listing as the file to view.

Exercise 22-11: Rewrite the source code from Listing 22-7 so that the filename can also be typed at the command prompt as the program's first argument.

TECHNICAL STUFF

- *Dump* is an old programming term. It's an inelegant way to refer to a transfer of data from one place to another without any manipulation. For example, a *core dump* is a copy of the operating system's kernel (or another basic component) transferred from memory into a file.
- The information saved by the *fwrite()* and read by the *fread()* functions is binary — effectively, the same information that's stored in memory when you assign a value to an *int* or a *float* or another C variable type.
- As long as you get the order correct, you can use *fwrite()* and *fread()* to save any data to a file, including full arrays, structures, and what-have-you. But if you read the information out of sequence, it turns into garbage.

Random File Access

Random file access has nothing to do with random numbers. Rather, the file can be accessed at any point hither, thither, and even yon. It's like watching streaming video as opposed to broadcast TV: You can start reading the file at any point, fetching or putting whatever information wherever you need it.

The best way to witness random file access in action is when the file is dotted with records of the same size. The notion of records brings up structures, which can easily be written to a file and then fetched back individually, selectively, or all at once.

Writing a structure to a file

As a type of variable, writing a structure to a file is cinchy. The process works just like writing any variable to a file, as demonstrated in Listing 22-8.

LISTING 22-8: Save Mr. Bond

```
#include <stdio.h>
#include <stdlib.h>
#include <string.h>

int main()
{
    struct agent {
        char actor[32];
        int year;
        char title[32];
    };
    struct agent bond;
    FILE *jbdb;

    strcpy(bond.actor,"Sean Connery");
    bond.year = 1962;
    strcpy(bond.title,"Dr. No");

    jbdb = fopen("bond.db","w");
    if(!jbdb)
    {
        puts("SPECTRE wins!");
        exit(1);
    }
    fwrite(&bond,sizeof(struct agent),1,jbdb);
    fclose(jbdb);
    puts("Record written");

    return(0);
}
```

Most of the code in Listing 22-8 should be familiar to you if you've worked through earlier exercises in this chapter.

Exercise 22-12: Copy the code from Listing 22-8 into your editor. Build and run the program to create the `bond.db` file, and write one structure to that file.

Exercise 22-13: Modify the code from Listing 22-8 so that a new program is created. Have that program write two more records to the `bond.db` file. They must be appended to any existing data, not overwriting the original file. Use this data:

```
Roger Moore, 1973, Live and Let Die
Pierce Brosnan, 1995, GoldenEye
```

Good information in a file doesn't do you any good unless you create code to read the data. Listing 22-9 reads in the three records written to the `bond.db` file, assuming that you've run program solutions to both Exercises 22-12 and 22-13.

LISTING 22-9: Get Me Bond!

```
#include <stdio.h>
#include <stdlib.h>
#include <string.h>

int main()
{
    struct agent {
        char actor[32];
        int year;
        char title[32];
    };
    struct agent bond;
    FILE *jbdb;
    int r;

    jbdb = fopen("bond.db","r");
    if(!jbdb)
    {
        puts("SPECTRE wins!");
        exit(1);
    }
    while( !feof(jbdb) )
    {
        r = fread(&bond,sizeof(struct agent),1,jbdb);
        if( r==0 )
            break;
```

```
        printf("%s\t%d\t%s\n",
                  bond.actor,
                  bond.year,
                  bond.title);
    }
    fclose(jbdb);

    return(0);
}
```

The source code in Listing 22-9 uses a *while* loop at Line 22 to read in the structures from the bond.db file. The code assumes that the file was created by writing full-size agent structures with the *fwrite()* function.

The *fread()* function returns the number of items read. It returns 0 if no data remains to be read. This value is tested in variable r at Line 25. If zero, the *while* loop is broken. Otherwise, the *feof()* function terminates the loop at Line 22.

The code uses the structure variable bond at Line 12 to read multiple items from a file. The new items overwrite any values already in the structure, just like reusing any variable.

Exercise 22-14: Create a new project by using the source code from Listing 22-9. Build and run to examine the bond.db file, which was created in Exercises 22-12 and 22-13.

TIP

The key to effectively read and write structures to a file — like a database — is to keep all structures uniform. This way, they can be read and written to the file in chunks. They can also be read or written in any order, as long as you also use the proper C language file functions.

Reading and rewinding

As your program reads data from a file, it keeps track of the position from whence data is read in the file. A file position index is maintained so that the location at which the code is reading or writing within a file isn't lost.

When you first open a file, this index position is located at the beginning of the file, the first byte. If you read a 40-byte record into memory, the file position index is 40 bytes from the start. If you read until the end of the file, the position index maintains that location as well.

To keep things confusing, the position index is often referred to as a *file pointer*, even though it's not a pointer variable or a *FILE* type of pointer. It's just the location within a file where the next byte of data is read.

You can mess with the file pointer-index-thing by using several interesting functions in C. Two of them are *ftell()* and *rewind()*. The *ftell()* function returns the current position as a *long int* value. The *rewind()* function returns the file pointer back to the start of the file.

Listing 22-10 reads twice through the records in the `bond.db` file. After reading all the records, the *rewind()* function at Line 37 resets the file pointer to the start of the file. The second *while* loop repeats the process, rereading all the records.

The *ftell()* function is used at Lines 26 and 42 to output the file pointer's offset each time the code reads records from the file.

LISTING 22-10: Tell and Rewind

```
#include <stdio.h>
#include <stdlib.h>
#include <string.h>

int main()
{
    struct agent {
        char actor[32];
        int year;
        char title[32];
    };
    struct agent bond;
    FILE *jbdb;
    int r;

    jbdb = fopen("bond.db","r");
    if(!jbdb)
    {
        puts("SPECTRE wins!");
        exit(1);
    }

    puts("First read-through:");
    while( !feof(jbdb) )
    {
        printf("%ld:\t",ftell(jbdb));
```

```
        r = fread(&bond,sizeof(struct agent),1,jbdb);
        if( r==0 )
            break;
        printf("%s\t%d\t%s\n",
                bond.actor,
                bond.year,
                bond.title);
    }

    /* restart! */
    rewind(jbdb);

    puts("Second read-through:");
    while( !feof(jbdb) )
    {
        printf("%ld:\t",ftell(jbdb));
        r = fread(&bond,sizeof(struct agent),1,jbdb);
        if( r==0 )
            break;
        printf("%s\t%d\t%s\n",
                bond.actor,
                bond.year,
                bond.title);
    }

    /* close and exit */
    fclose(jbdb);
    return(0);
}
```

Exercise 22-15: Type the source code from Listing 22-10 into your editor. Build and run to see how the *ftell()* and *rewind()* functions operate.

Finding a specific record

When a file contains records all of the same size, such as the James Bond database used so far in this chapter, you can use the *fseek()* function to pluck out any individual item. This is the essence of random file access. The format for *fseek()* is

```
fseek(handle,offset,whence);
```

`handle` is a file handle, a *FILE* pointer representing a file that's open for reading. `offset` is the number of bytes from the start, end, or current position in a file. And `whence` is one of three constants: SEEK_SET, SEEK_CUR, or SEEK_END for the start, current position, or end of a file, respectively.

Providing that the data file contains records of a constant size, you can use *fseek()* to pluck out any specific record, as shown in Listing 22-11.

LISTING 22-11: Find a Specific Record in a File

```
#include <stdio.h>
#include <stdlib.h>
#include <string.h>

int main()
{
    struct agent {
        char actor[32];
        int year;
        char title[32];
    };
    struct agent bond;
    FILE *jbdb;

    jbdb = fopen("bond.db","r");
    if(!jbdb)
    {
        puts("SPECTRE wins!");
        exit(1);
    }
    /* locate the 2nd record based on the
       size of structure agent */
    fseek(jbdb,sizeof(struct agent)*1,SEEK_SET);
    fread(&bond,sizeof(struct agent),1,jbdb);
    printf("%s\t%d\t%s\n",
            bond.actor,
            bond.year,
            bond.title);
    fclose(jbdb);

    return(0);
}
```

The *fseek()* function at Line 21 sets the file-pointer-thing position so that the *fread()* function that follows (Line 22) reads in a specific record located inside the database. The offset is calculated by multiplying the size of the `entry` structure. As with an array, multiplying this size by 1 yields the *second* record in the file; multiply the value by 0 (or just specify 0 in the function) to read the first record. The `SEEK_SET` constant ensures that *fseek()* looks from the beginning of the file.

Exercise 22-16: Using the source code from Listing 22-11, create a new program to see the second record in the file. This exercise's success depends on the existence of the `bond.db` file, which has been built throughout this chapter.

Saving a linked list to a file

Chapter 20 ponders the ponderous topic of linked lists in C. One question that inevitably surfaces during the linked-list discussion is how to save such a list to a file. If you've read the past few sections, you already know how: Create the file and then use *fwrite()* to save all the linked-list records.

Exercise 22-17: Modify the source code from Exercise 20-12 (refer to Chapter 20) so that the program retrieves and saves all records to a file. The program should automatically load the records when it starts. As it quits, the program automatically saves the records. The code needs two new functions, *load()* and *save()*, which you can base upon the existing *create()* and *show()* functions, respectively. Of course, other spiffing-up is required, as usual.

Here are some pointers [sic] for creating Exercise 22-17:

TIP

» It's inevitable that saving a linked-list structure saves a pointer's address. Doing so is okay — just discard the value when you read in the structure, replacing it with the new address allocated for the linked structure.

» When the code runs the first time, obviously it won't find a file to load into memory. That's okay; have your program create the file.

» As with any complex coding, tackle the improvements one step at a time.

» Checking file types

» Working with the directory hierarchy

» Changing filenames

» Duplicating files

» Removing a file

Chapter 23
File Management

The C library features many functions that interface directly with the operating system, allowing you to peek, poke, and prod into the very essence of files themselves. You never know when you need to plow through a directory, rename a file, or delete a temporary file that the program created. It's powerful stuff, but such file management is well within the capabilities of your C programs.

Directory Madness

A *directory* is nothing more than a database of files stored on a device's mass storage system. Also called a *folder*, a directory contains a list of files plus any subdirectories. Just as you can manipulate a file, a directory can be opened, read, and then closed. And, as with the directory listing you see on a computer screen, you can gather information about the various files, their sizes, types, and more.

Calling up a directory

The C library's *opendir()* function examines the contents of a specific directory. It works similarly to the *fopen()* function. Here's the format:

```
dhandle = opendir(pathname);
```

dhandle is a pointer of the *DIR* type, like a file handle being of the *FILE* type. The *pathname* is the name of a directory to examine. It can be a full path, or you can use the . (dot) abbreviation for the current directory or .. (dot-dot) for the parent directory.

Once a directory is open, the *readdir()* function fetches records from its database, similar to the *fread()* function, though the records describe files stored in the directory. Here's the *readdir()* function's format:

```
*entry = readdir(dhandle);
```

entry is a pointer to a `dirent` structure. After a successful call to *readdir()*, the structure is filled with information about a file in the directory. Each time *readdir()* is called, it points to the next file entry, like reading records from a database. When the function returns NULL, the last file in the directory has been read.

Finally, after the program is done messing around, the directory must be closed. This operation is handled by the *closedir()* function:

```
closedir(dhandle);
```

All these directory functions require the `dirent.h` header file to be included with your source code.

Listing 23-1 illustrates code that reads a single entry from the current directory. Two required variables are declared in Lines 7 and 8: `folder` is a *DIR* pointer, used as the handle to represent the directory that's opened, and `file` is the memory location of a structure that holds information about individual files in the directory.

LISTING 23-1: Pluck a File from the Directory

```
#include <stdio.h>
#include <stdlib.h>
#include <dirent.h>

int main()
{
    DIR *folder;
    struct dirent *file;

    folder=opendir(".");
    if(folder==NULL)
    {
        puts("Unable to read directory");
        exit(1);
```

```
    }
    file = readdir(folder);
    printf("Found the file '%s'\n",file->d_name);
    closedir(folder);
    return(0);
}
```

The directory is opened at Line 10; the single dot is an abbreviation for the current directory. Lines 11 through 15 handle any errors, similar to opening any file. (Refer to Chapter 22.)

The first entry in the directory is read at Line 16, and then Line 17 displays the information. The `d_name` member of the `dirent` structure references the file's name.

At Line 18, the directory is closed.

Exercise 23-1: Create a new project by using the source code from Listing 23-1. Build and run.

Of course, the first file that's most likely to be read in a directory is the directory itself, the dot entry. Boring!

Exercise 23-2: Modify the source code shown in Listing 23-1 so that the entire directory is read. A *while* loop can handle the job.

REMEMBER

The *readdir()* function returns NULL after the last file entry has been read from a directory. In fact, the expression `file=readdir(folder)` evaluates to NULL when the last entry has been read from a directory.

Gathering more file info

The *stat()* function reads various and sundry information about a file: its date, size, type, and other trivia. The function's format looks like this:

```
stat(filename,stat);
```

`filename` is a string value, the name of the file to examine. *`stat`* is the address of a `stat` structure. After a successful call to the *stat()* function, the `stat` structure is filled with information about the file. And I wholly agree that calling both the function and the structure *stat* leads to an undue amount of consternation.

You need to include the `sys/stat.h` header file in your code to make the compiler pleased with the *stat()* function.

Listing 23-2 demonstrates how the *stat()* function can be incorporated into a directory listing. It starts with the inclusion of the `sys/stat.h` header file at Line 5. The `sys/` part directs the compiler to look in that directory to locate the `stat.h` file. (`sys` is a subdirectory of `include`, where C header files are stored.)

LISTING 23-2: A More Impressive File Listing

```
#include <stdio.h>
#include <stdlib.h>
#include <dirent.h>
#include <time.h>
#include <sys/stat.h>

int main()
{
    DIR *folder;
    struct dirent *file;
    struct stat filestat;

    folder=opendir(".");
    if(folder==NULL)
    {
        puts("Unable to read directory");
        exit(1);
    }
    while( (file = readdir(folder)) != NULL )
    {
        stat(file->d_name,&filestat);
        printf("%-14s %5ld %s",
            file->d_name,
            filestat.st_size,
            ctime(&filestat.st_mtime));
    }
    closedir(folder);
    return(0);
}
```

Line 11 creates a `stat` structure variable named `filestat`. This structure is filled at Line 21 for each file found in the directory; the `file->d_name` member provides the filename as the *stat()* function's first argument. The second argument is the address of the `filestat` structure.

The *printf()* function starting at Line 22 outputs the information revealed by the *stat()* function: Line 23 displays the file's name; Line 24 pulls the file's size from the `filestat` structure; and in Line 25, the *ctime()* function extract's the file's modification time from the `filestat` structure's `st_mtime` member. (Refer to Chapter 21 for more information about time programming in C.)

Oh! And the *printf()* statement lacks a `\n` (newline) because the *ctime()* function's output provides one.

Exercise 23-3: Type the source code from Listing 23-2 into your editor, or just modify your solution from Exercise 23-2. Build and run to see a better directory listing.

Separating files from directories

Each file stored in a directory is classified by a file type or mode. For example, some entries in a directory listing are subdirectories. Other entries may be symbolic links or sockets. To determine which file is of which type, your code can examine the `st_mode` member of the `stat` structure.

The `st_mode` value is a *bit field* — bits in this value are set depending on the various file type attributes and permissions applied to the file. To examine the bits, you use special macros made available in the `dirent.h` header.

For example, the `S_ISDIR` macro returns TRUE when a file's `st_mode` element indicates a directory, not a regular file. Use the `S_ISDIR` macro like this:

```
S_ISDIR(filestat.st_mode)
```

This expression evaluates TRUE for a directory, and FALSE otherwise.

Exercise 23-4: Modify your solution to Exercise 23-3 so that any subdirectories listed are flagged as such. Because directories don't generally have file sizes, specify the text `<DIR>` in the file size field for the program's output.

TIP

If the current directory lacks subdirectories, change the directory name in Line 13.

In Windows, use two backslashes when typing a path. For example:

```
dhandle = opendir("\\Users\\Dan");
```

Windows uses the backslash as a pathname separator. C uses the backslash as an escape character in a string. To specify a single backslash, you must specify two of them.

Exploring the directory tree

Most storage media feature more than one directory. The main directory is the root, with subdirectories organizing the media. Using C, you can create directories of your own and flit between them like bees upon flowers. The C library sports various functions to sate your directory-diving desires. Here's a sampling:

getcwd()	Retrieve the current working directory
mkdir()	Create a new directory
chdir()	Change to the directory specified
rmdir()	Obliterate the directory specified

getcwd(), *chdir()*, and *rmdir()* require the `unistd.h` header file; the *mkdir()* function requires `sys/stat.h`.

Listing 23-3 makes use of three directory functions: *getcwd()*, *mkdir()*, and *chdir()*.

LISTING 23-3: Make Me a Directory

```
#include <stdio.h>
#include <unistd.h>
#include <sys/stat.h>

int main()
{
    char curdir[255];

    getcwd(curdir,255);
    printf("Current directory is %s\n",curdir);
    mkdir("very_temporary",S_IRWXU);
    puts("New directory created.");
    chdir("very_temporary");
    getcwd(curdir,255);
    printf("Current directory is %s\n",curdir);
    return(0);
```

The *getcwd()* function in Line 9 captures the current directory's name and saves it in the `curdir` array. The directory name — a full pathname — is output at Line 10.

Line 11 creates a new directory, `very_temporary`. The defined constant `S_IRWXU` is the file-creation mode. This constant sets the new directory's permissions, allowing the owner read, write, and execute access.

WARNING

If you're compiling for Windows, you must omit the second argument for the *mkdir()* function at Line 11:

```
mkdir("very_temporary");
```

After the directory is created, the *chdir()* function on Line 13 changes to that directory, followed by the *getcwd()* function at Line 14 capturing its full pathname.

When the program is done, its environment is purged, which means the directory in which the program is running is restored; the program changes to the new directory only while the program is running.

Exercise 23-5: Copy the source code from Listing 23-3 into your editor. Remember to omit the second argument for *mkdir()* at Line 11 if you're compiling on Windows. Build and run the program.

After running the program for Exercise 23-5, a new directory, `very_temporary`, is created in whichever directory the program was run. Feel free to remove that directory using your computer operating system's directory obliteration command.

TECHNICAL STUFF

In Listing 23-3, Line 7 sets aside 255 characters for storing the current directory's pathname. I'm plucking the value `255` out of thin air; it should be large enough. Serious programmers use a constant defined for their systems. For example, `PATH_MAX` defined in the `sys/syslimit.h` header file is perfect, but it's not available on all systems. You could use the `FILENAME_MAX` constant (defined in `stdio.h`), but it sets the size for a filename, not a full pathname. As a compromise, I choose `255`.

Fun with Files

The C library offers functions for making a new file, writing to that file, and reading data from any file. Bolstering these file I/O functions are a suite of file manipulation functions. They allow your programs to rename, copy, and delete files. The functions work on any file, not just those you create, so be careful!

Renaming a file

The *rename()* function is not only appropriately named but also easy to figure out:

```
x = rename(oldname,newname);
```

`oldname` is the name of a file already present; `newname` is the file's new name. Both values can be immediate or variables. The return value is 0 upon success, and −1 otherwise.

The `rename()` function is prototyped in the `stdio.h` header file.

The source code shown in Listing 23-4 creates a file named `blorfus` and then renames that file to `wambooli`.

LISTING 23-4: Creating and Renaming a File

```
#include <stdio.h>
#include <stdlib.h>

int main()
{
    FILE *test;

    test=fopen("blorfus","w");
    if(!test)
    {
        puts("Unable to create file");
        exit(1);
    }
    fclose(test);
    puts("File created");
    if(rename("blorfus","wambooli") == -1)
    {
        puts("Unable to rename file");
        exit(1);
    }
    puts("File renamed");
    return(0);
}
```

Lines 8 through 14 create the file `blorfus`. The file is empty; nothing is written to it.

The *rename()* function at Line 16 renames the file. The return value is compared with −1 to confirm that the operation is successful.

Exercise 23-6: Create a new program by using the source code shown in Listing 23-4. Build and run.

The renamed file, wambooli, is used in a later section as an example.

Copying a file

The C library features no function that duplicates a file. Instead, you must craft your own: Write code that reads in a file, one chunk at a time, and then writes out that chunk to a duplicate file. This method is how files are copied.

Listing 23-5 demonstrates how a file can be duplicated, or copied. The two files are specified in Lines 9 and 10. In fact, Line 9 uses the name of the exercise file, the source code from Listing 23-5. The destination file, which contains the copy, is the same filename, but with a bak extension.

LISTING 23-5: Duplicate That File

```
#include <stdio.h>
#include <stdlib.h>

int main()
{
    FILE *original,*copy;
    int c;

    original=fopen("ex2307.c","r");
    copy=fopen("ex2307.bak","w");
    if( !original || !copy)
    {
        puts("File error!");
        exit(1);
    }
    while( (c=fgetc(original)) != EOF)
        fputc(c,copy);
    puts("File duplicated");
    return(0);
}
```

The copying work is done by the *while* loop at Line 16. One character is read by the *fgetc()* function, and it's immediately copied to the destination by the *fputc()* function in Line 17. The loop keeps spinning until the EOF, or end-of-file, is encountered.

Exercise 23-7: Copy the source code from Listing 23-5 into your editor. Save the file as ex2307.c (which is this book's file-naming convention), build, and run. You need to use your computer's operating system to view the resulting file in a folder window. Or, for *For Dummies* bonus points, you can view the results in a terminal or command prompt window.

Deleting a file

Programs delete files all the time, though these files are mostly temporary anyway. Back in the bad old days, I remember complaining about programs that didn't "clean up their mess." If your code creates temporary files, remember to remove them as the program quits. The way to do that is via the *unlink()* function.

Yes, the function is named *unlink* and not *delete* or *remove* or *erase* or whatever nomenclature you're otherwise used to. In Unix, the *unlink* command is used in the terminal window to zap files, though the *rm* command is more popular.

The *unlink()* function requires the presence of the unistd.h header file, which you see at Line 3 in Listing 23-6.

LISTING 23-6: File Be Gone!

```
#include <stdio.h>
#include <stdlib.h>
#include <unistd.h>

int main()
{
    if(unlink("wambooli") == -1)
    {
        puts("I just can't kill that file");
        exit(1);
    }
    puts("File killed");
    return(0);
}
```

The file slated for death is listed in Line 7 as the *unlink()* function's only argument. It's the wambooli file, created back in Exercise 23-6! So, if you don't have that file, go back and work Exercise 23-6. (In Code::Blocks, you must copy this file into the proper folder for your solution to Exercise 23-8.)

Exercise 23-8: Type the source code from Listing 23-6 into your editor. Build and run.

» Combining multiple source code files

» Making your own header file

» Linking in additional libraries

Chapter 24

Beyond Mere Mortal Projects

Not every C program you write will have only 20 or 30 lines of code. Most of the programs, the ones that truly do something, are much longer. Much, much longer. Some become so huge that it makes sense to split them into smaller modules, or individual source code files, with maybe 20 to 60 lines of code apiece. Not only do these smaller modules make it easier to write and update code, but you can also reuse common modules in other projects, reducing development time.

The Multi-Module Monster

The C language places no limit on how long a source code file can be. Likewise, a source code file can consist of only a few lines — if you can pull off this trick. The determination of whether to use multiple source code files — *modules* — really depends on the programmer. That's you. How easy do you want to make the process of writing, maintaining, and updating your code?

Linking two source code files

The most basic multi-module monster project has two source code files. Each file is separate — written, saved, and compiled individually — but eventually brought together as one unit by the linker. The *linker*, which is part of the build process in Code::Blocks, is what creates a single program from several different modules.

What's a module?

A *module* is a compiled object file. The various object files are linked to build a program. The entire operation starts with separate source code files, starting with Listing 24-1.

LISTING 24-1: The *main.c* Source Code File

```
#include <stdio.h>
#include <stdlib.h>

void second(void);

int main()
{
    printf("Second module, I send you greetings!\n");
    second();
    return 0;
}
```

Listing 24-1 shows the `main.c` module. In it, you see the *second()* function prototyped at Line 4. This prototype is required because the *second()* function is called at Line 11, but it exists in another module — just like C library functions exist elsewhere. You don't need to prototype all functions found in another module, only those referenced or called.

Exercise 24-1: Fire up a new project in your IDE named `ex2401`. Create the project as you normally would: Type the source code from Listing 24-1 into the editor as the contents of the `main.c` file. Save the file.

Don't build yet! After all, the code references the *second()* function, which doesn't seem to exist anywhere. It's prototyped, as is required for any function that's used in your code, but the *second()* function is found in another module. To create this next module in the Code::Blocks IDE, follow these steps:

1. **Save the current project.**
2. **Choose File ➪ New ➪ Empty File.**
3. **Click the Yes button when you're prompted to add the file to the active project.**

   The Save File dialog box appears.

4. **Type** alpha.c **as the filename and then click the Save button.**

   The new file is listed on the left side of the Code::Blocks window, beneath the Sources heading where the `main.c` file is listed. A new tab appears in the editor window, with the `alpha.c` file ready for editing, as shown in Figure 24-1.

5. **Edit the new source code file.**

   Click its tab, if necessary. (Refer to Figure 24-1.)

6. **Type the source code from Listing 24-2 into the** `alpha.c` **file in Code::Blocks.**
7. **Save the project.**
8. **Build and run.**

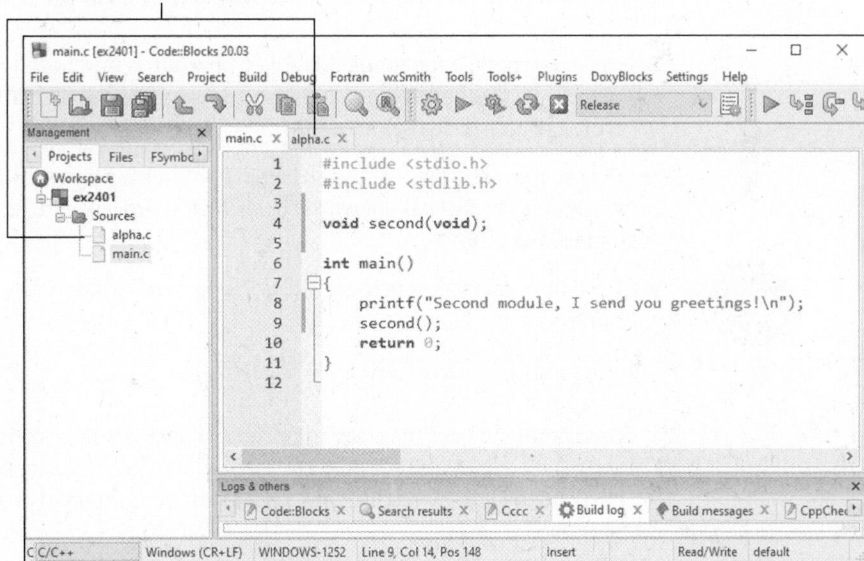

FIGURE 24-1: Two source code files in the Code::Blocks IDE.

LISTING 24-2: The *alpha.c* Source Code File

```
#include <stdio.h>

void second(void)
{
    puts("Glad to be here!");
}
```

The `alpha.c` module contains the *second()* function called from the first module. Because both modules are linked as the program is built, the function is found when it's called.

Here's the output I see on my computer:

```
Second module, I send you greetings!
Glad to be here!
```

The two source code files aren't "glued together" by the compiler; each source code file is compiled individually. A separate object code file is created for each one: `main.o` and `alpha.o`. It's these two object code files that are then linked together, combined with the C standard library, to form the final program.

TECHNICAL STUFF

- The main module for a multi-module C program is traditionally named `main.c`. This is the reason that Code::Blocks names the first (and, often, only) project source code file `main.c`.
- Only source code files contained within the same project — found beneath the Sources branch, as shown in Figure 24-1 — are linked together in the Code::Blocks IDE.
- To compile and link source code files in a terminal window, use the following command:

  ```
  clang main.c alpha.c -o ex2401
  ```

  This command uses the *clang* compiler to build the source code files `main.c` and `alpha.c` into a single project. The output switch (-o) sets the program filename to `ex2401`. To run, type **./ex2401** at the command prompt.

Sharing variables between modules

The best way to share a variable between several functions in a huge project is to make that variable external. (The specifics for this operation are found in

Chapter 16.) An external variable, also called a global variable, must be declared in only one module, usually the main module. This variable is declared outside of any function, usually at the top of the source code file, betwixt the `#include` directives and function prototypes. It works like any variable declaration:

```
type name;
```

with *type* being the data type and *name* the assigned name, optionally followed by an assignment or initialization. To access this external variable from other modules, you must employ the *extern* keyword. This keyword doesn't declare the variable, but rather references that it's declared elsewhere, in another source code module. Here's the *extern* keyword's format:

```
extern type name;
```

Listing 24-3 shows an example of how an external variable is declared, with Listing 24-4 illustrating how the *extern* keyword is put to work.

LISTING 24-3: Code for *main.c* and a Global Variable

```
#include <stdio.h>
#include <stdlib.h>

void second(void);

int count;

int main()
{
    for(count=0;count<5;count++)
        second();
    return 0;
}
```

External variable `count` is declared at Line 6. It's used in the *for* loop at Line 10, but it's also used in the `second.c` source code file, shown in Listing 24-4.

LISTING 24-4: Code for *second.c* Using the Global Variable

```
#include <stdio.h>

extern int count;

void second(void)
{
    printf("%d\n",count+1);
}
```

The second.c source code file declares external variable count, which is found in the main.c file. To properly access this variable, Line 3 in Listing 24-4 identifies it as an external *int*. The count variable is then used in the *second()* function — specifically, at Line 7. Its value is retained between modules and all functions in which it's used.

Exercise 24-2: Create a new project in Code::Blocks that incorporates both source code files shown in Listings 24-3 and 24-4. Build and run.

To build the project at the command prompt: Update the two source code files from Exercise 24-1 and build per the directions at the end of the preceding section.

Creating a custom header file

As multi-module projects grow more complex, you find that the first part of each source code file gets longer and longer: More prototypes, more constants, and more external variables, structures, and other whatnot are required for each module. Rather than burden your source code files with redundancies, you can create a header file for the project.

A header file contains just about everything you can put into a source code file. Specifically, it's best to put items in the header file that would otherwise be required for every source code module. Listing 24-5 shows a sample header file.

LISTING 24-5: Header File *ex2403.h*

```
#include <stdio.h>
#include <stdlib.h>

/* structures */
struct thing {
```

```
    char name[32];
    int age;
    };
typedef struct thing human;

/* prototypes */

void fillstructure(void);
void printstructure(void);

/* defined constants */

/* variables */
```

The header file shown in Listing 24-5 starts with some `#include` directives, which works well when these header files are required by each module in the program.

The structure `thing` is defined at Line 5, with Line 8 specifying a *typedef* so that the word *human* (instead of `struct thing`) can be used in the code.

REMEMBER

Structures must be declared before any functions or prototypes that use the structure.

Next come two prototypes at Lines 13 and 14. These prototypes save you the bother of prototyping each function referenced in the various modules.

The header file in Listing 24-5 lacks defined constants and external variables, which could easily be placed there.

Other popular items to include in a header file are macros. These are preprocessor directives that can also help simplify your code. You can read about them at my blog:

```
c-for-dummies.com/blog/?page_id=2
```

WARNING

One thing not found in traditional C header files is a function. Putting a function into a header is something that C++ programmers may do, but it's unusual in C. Keep your functions in the source code files, where they're expected. Do not set them into header files.

To use a local header file, you specify the `#include` preprocessor directive at the start of the source code file, as in any other header file. The big difference is that double quotes are used instead of angle brackets. For example:

```
#include "ex2403.h"
```

The double quotes direct the compiler to look for the header file in the current directory. If the file isn't in that directory, you must specify a pathname, as in

```
#include "headers/ex2403.h"
```

Listing 24-6 uses the header file from Listing 24-5.

LISTING 24-6: Project ex2403 *main.c* Source Code

```
#include "ex2403.h"

human person;

int main()
{
    fillstructure();
    printstructure();
    return 0;
}

void fillstructure(void)
{
    printf("Enter your name: ");
    fgets(person.name,31,stdin);
    printf("Enter your age: ");
    scanf("%d",&person.age);
}

void printstructure(void)
{
    printf("You are %s\n",person.name);
    printf("And you are %d years old.\n",person.age);
}
```

Line 1 of the source code shown in Listing 24-6 includes the custom header file, `ex2403.h`. The *typedef* `human` is then used at Line 3 to declare structure variable `person`. That's it! No other declarations are necessary in the source code, because they've been handled by the custom header.

Exercise 24-3: Create a new project in Code::Blocks. Write a new header file, `ex2403.h`, for the project using code from Listing 24-5. Use the steps from the earlier section "Linking two source code files" to create a new file, naming it `ex2403.h` and adding it to the current project. Copy the source code from Listing 24-6 into the `main.c` file. Build and run.

REMEMBER

Ensure that the header file `ex2403.h` is in the same build folder as the `main.c` file when building the project in Code::Blocks. If it isn't, the compiler belches up a fatal error.

TIP

You don't specify the header file when compiling at the command line. The command to build Exercise 24-3 is **clang main.c -o main**, where the program output is named `main`. The compiler stirs in the `ex2403.h` header file as directed to do so in the `main.c` source code file.

Exercise 24-4: Split out the *fillstructure()* and *printstructure()* functions from Listing 24-6 so that each appears in its own source code module, `input.c` and `output.c` for filenames, respectively. Build the multi-module program.

TIP

THOUGHTS ON SPLITTING UP CODE

I divide large program modules by function. For example, all output functions go into a `display.c` module; input functions belong in `input.c`. I create an `init.c` module for initialization routines. Beyond that, the number of modules depends on what the program does.

Putting similar functions into a module is a good idea, though having one function in a module is also okay. In fact, when you do, and the module works, you can pretty much set it aside when it comes to working out bugs and whatnot.

No matter what the project size, I recommend creating a project header file. That header file keeps all function prototypes, global variables, and constants in one place — plus, it helps map out the entire project. For example, you can list function prototypes by module or add comments, development history, and other notes.

Other Libraries to Link

Throughout this book, your programs have linked in the standard C library. This process works automatically. It's all that's needed for the basic console applications in this book. When your programs require more sophistication, however, they can link in other libraries.

If you're coding something graphical, for example, you can link in a graphics library. Or, if you're doing fancy console (text) programming, you can link in the NCurses library. These libraries, and the functions they include, greatly augment a program's capabilities.

A function's documentation helps determine which libraries to use. For example, in Linux the *pow()* function's *man* page has this note right up front:

```
Link with -lm.
```

This direction tells you to use the compiler's command line `-l` (little *L*) switch immediately followed by `m` (little *M*) for the math library. This switch directs the linker to bring in the math library, without which you'd witness a flood of errors for unknown functions, like *pow()*.

Linking in a library is different from including a header file. The header file is handled by the `#include` directive. It contains prototypes, definitions, and other helpful information. But it's the library that contains the gears that make the functions work.

All C compilers link in the standard C library, called `c`. If another library needs to be linked in, it's specified directly. The library must be available (on the local storage system), and you must know its location when using an IDE.

For example, in Code::Blocks, a library is added by choosing Project ⇨ Build Options and adding the library on the Linker Settings tab. You must know the library's folder and its name to complete this task.

At the command prompt, the `-l` (little *L*) switch is immediately followed by the library's name. This switch might need to be specified as the last argument with certain compilers.

The variety and purpose of the various libraries available to a C compiler depend on what you plan to do. C language library packages can be found for free and downloaded from the Internet. For example, a hardware manufacturer may provide a library you can use to program its specific device.

- Libraries are linked in only when creating a Code::Blocks project. You cannot link in a library for an individual source code file.
- Traditionally, C libraries are stored in the `/usr/lib` folder, though downloaded libraries might be found in `/usr/local/lib` or elsewhere. Command-line compilers automatically look in these locations.
- The -L switch for a command-line compiler can be specified to direct the compiler to look in a specific folder for a library. For example:

```
clang -L/usr/local/share/lib -lcompress main.c
```

  The preceding command directs the *clang* compiler to look in the `/usr/local/share/lib` folder for the `compress` library.

» **Configuring a project for debugging**

» **Using Code::Blocks' GNU debugger**

» **Stepping through a program**

» **Checking the values of variables**

» **Writing better error messages**

Chapter 25

Out, Bugs!

Everyone writes buggy code. Not intentionally, of course. Even the best programmers get lazy or sloppy and do silly things. Stray semicolons, misplaced commas, and missing curly brackets happen to all programmers. Fortunately, the compiler catches a lot of this crummy code. Fix the source code and recompile to deal with those annoying, typical bugs.

For deeper problems, flaws in logic, or maybe code boo-boos that aren't easy to find, it helps to have a little assistance. That assistance comes in the form of debugging your code, which you can do manually or with the help of a debugger. The goal is to see what's gone wrong.

Simple Tricks to Resolve Problems

When I can't figure out what's going on with a program and I'm too lazy to run it through a debugger, I use the *printf()* and *puts()* functions as my debugging friends. These tools can output helpful messages and values, assisting me in determining what's buggy with the code.

Documenting the flow

Suppose a function receives variable x but, somehow, variable x never shows up. Where did it go? To find out, I insert the following statement into the code:

```
printf("value of 'x' at Line 125: %d\n",x);
```

This statement may appear in several places, tracing the value of variable x as it moves through the code. The output clues me in to where variable x wanders off. Further, you can use the `__LINE__` macro instead of hard-coding the line number:

```
printf("value of 'x' at Line %d: %d\n",__LINE__,x);
```

The preprocessor expands the `__LINE__` macro — and yes, it has two underscores before and after — into the current source code line number. By using this trick, you can set an output message and continue to edit your code without manually updating line numbers.

When I'm not tracking a variable and I only want to know why a chunk of code isn't executing, I insert a *puts()* statement, like this:

```
puts("You got to here");
```

When I see the preceding text in the output, I know that the code is being approached but still may not be executed. At this point, I examine the code nearby, looking for common C language boo-boos, many of which are documented in Chapter 26.

Talking through your code

An excellent way to catch flaws in program flow is to talk through your code. The catch is that you must do so out loud. Yes, this trick can be embarrassing, but it's highly effective.

For example, some code isn't working and you have no clue why. Pretend you're speaking with a fellow programmer. Explain to the phantom person what your code is doing and how it works. As you talk through your code, the problem may suddenly appear. Believe it or not, this trick has worked for me dozens of times.

Writing comments for future-you

Another thing you can do to help fix undue woe is to use comments to describe the problem in the code. This approach may not fix the problem now, but for future-you looking at the code down the line, it's a real help; it beats trying to discover the boo-boo all over again.

For example:

```
for(y=x+a;y<c;y++)      /* this doesn't seem to work */
    manipulate(y);      /* Confirm that 'a' is changing */
```

In this example, the note reminds future-me that the statements aren't doing what they're intended; plus, it offers future-me a suggestion on what to look for in a solution.

TIP

You can also use comments to offer future-you suggestions on how to improve the code, things to tighten up, or new features you just don't have time to add.

The Debugger

Many popular debuggers exist, tools that let you examine your code line-by-line as it runs, peeking at memory and looking at variables as they change. The popular GNU debugger is available at the command prompt — if you're bold enough to use it. For mere mortals, consider using the Code::Blocks debugger.

Debugging setup

To debug a project, you need to set its target — the program — to include debugging information. The debugger uses this information to help process your code and to see how things run — or not. This process works only with a Code::Blocks project and only when you create a debugging target build for your code. Follow these steps:

1. **Start a new project in Code::Blocks.**

   Choose File ➪ New ➪ Project.

2. **Choose Console Application and click Go.**
3. **Choose C and click Next.**
4. **Type the project title, such as** ex2501 **for Exercise 25-1.**
5. **Click the Next button.**

   So far, these first few steps are the same as for creating any C language console program in Code::Blocks.

6. **Place a check mark by the option Create "Debug" Configuration.**

   This setting directs Code::Blocks to include debugging information when the program is built.

7. **Ensure that the item Create "Release" Configuration is also selected.**
8. **Click the Finish button.**

   The new project appears in Code::Blocks.

When you activate debugging for a project, as well as keeping the release configuration (refer to Step 7), you can use the Compiler toolbar to choose which version of the code is created, as shown in Figure 25-1. Use the View⇨ Toolbars⇨ Compiler command to show or hide that toolbar.

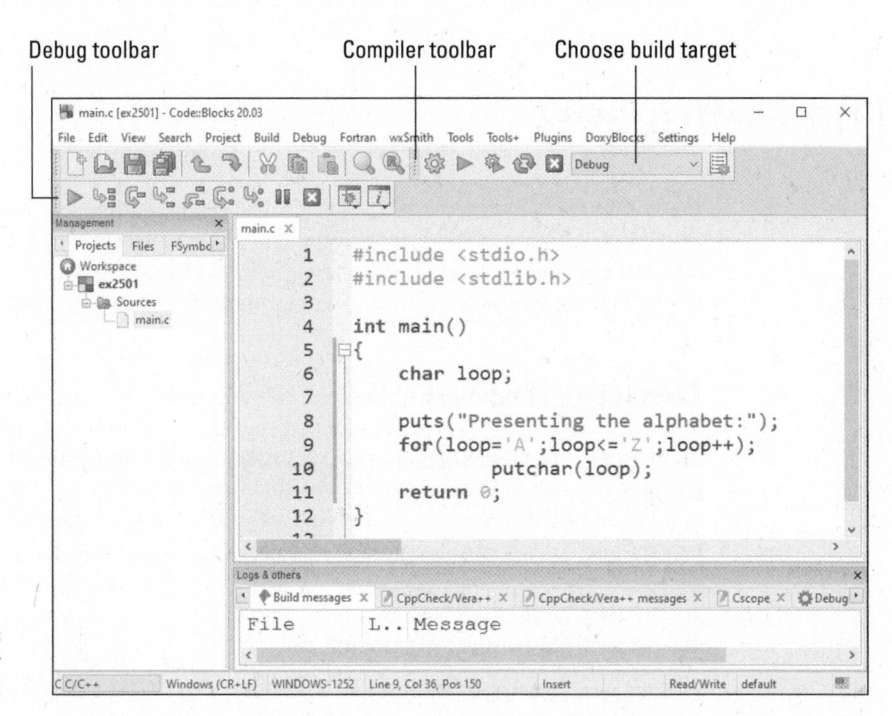

FIGURE 25-1: The Compiler toolbar.

When debugging, ensure that the Debug command is chosen as the build target. You cannot debug the code unless the debugging information is included in the final program.

REMEMBER

To create the final program when you're finished debugging, choose the Release command from the Build Target menu. Though you could release a debugging version of your program, this information makes the final program larger.

Working the debugger

The debugger operates by examining your code as it runs, showing you what's happening, both internally to the program as well as the output. If you've created a new Code::Blocks program with debugging information (see the preceding section) and you have code to debug, you're ready to start.

I confess that the code shown in Listing 25-1 is purposefully riddled with bugs.

LISTING 25-1: Debug Me!

```
#include <stdio.h>
#include <stdlib.h>

int main()
{
    char loop;

    puts("Presenting the alphabet:");
    for(loop='A';loop<='Z';loop++);
        putchar(loop);
    return 0;
}
```

Exercise 25-1: Create a new project in Code::Blocks, one that has a Debug target build. Copy the source code from Listing 25-1 into the `main.c` file. Ensure that you copy the text exactly, including a mistake you may see at the end of Line 9. Build and run.

Because the Code::Blocks editor is smart, as are other programming editors, you may catch the erroneous semicolon at the end of Line 9 because the following line didn't automatically indent. The compiler may also catch the boo-boo. These are big clues, but also things you may not notice, especially if you have 200 lines of code to look at. Regardless, the program's output tells you something is amiss. Here's what I see:

```
Presenting the alphabet:
[
```

The alphabet doesn't show up, of course. Not only that, what's the `[` character for? Time to debug!

Use the Debugger toolbar in Code::Blocks to help you wade into your code to see what's fouled up. (Refer to Figure 25-1.) To show or hide that toolbar, choose View ⇨ Toolbars ⇨ Debugger.

Follow these steps to work through your code to determine what's wrong:

1. **Click the cursor in your code right before the *puts()* statement.**

   This location is at Line 8.

2. **Click the Run to Cursor button on the Debugging toolbar.**

   The Run to Cursor button is illustrated in the margin.

   The program runs, but only up to the cursor's location. The output window appears, and debugging information shows up on the Debugging tab at the bottom of the Code::Blocks window.

3. **Click the Next Line button.**

   The *puts()* statement executes, and its output appears in the output window.

4. **Click the Next Line button again.**

   The *for* loop does its thing; no output.

5. **Click the Next Line button again.**

   The *putchar()* function outputs the character lurking in the `loop` variable. Because the loop already went through ASCII A to Z, the next character is [. This is the character output.

   Hopefully, at this point you look closer at your code and find the stray semicolon at the end of Line 9.

6. **Click the Stop button to halt the debugger.**
7. **Remove the semicolon at the end of Line 9.**
8. **Save and rebuild your code.**

   To determine whether you've fixed the problem, step through the code again:

9. **Click the mouse pointer to place the cursor right before the *for* statement at Line 9.**
10. **Click the Run to Cursor button.**
11. **Click the Next Line button twice.**

    An *A* appears as output. Good.

12. **Keep clicking the Next Line button to work through the *for* loop.**

    When you're satisfied that the code has been debugged:

13. **Click the Stop button.**

The program runs fine after you fix the stray semicolon. See the later section "Watching variables" for insight into how variable values can be examined by the debugger.

Setting a breakpoint

No one wants to step through 200 lines of source code to find a bug. Odds are that you have a good idea where the bug is, either by the program's output or because it ran just five minutes ago, before you edited that one particular section. If so, you know where you want to snoop into operations. It's at that place in your code that you set a debugging breakpoint.

A *breakpoint* is like a stop sign in your text. In fact, that's the exact icon used by Code::Blocks, as shown in Figure 25-2. To set a breakpoint, click the mouse between the line number and the green line (or yellow line, if you haven't saved yet). The Breakpoint icon appears.

Breakpoint

main.c

```
1   #include <stdio.h>
2   #include <stdlib.h>
3
4   int main()
5   {
6       char loop;
7
8       puts("Presenting the alphabet:");
9       for(loop='A';loop<='Z';loop++)
10              putchar(loop);
11      return 0;
12  }
```

FIGURE 25-2: A breakpoint in the code.

To run your code to the breakpoint, click the Debug/Continue button on the Debugging toolbar. The program runs, but then comes to a screeching halt at the breakpoint. From this point on, you can work the debugger to see what's going wrong.

Watching variables

Sometimes, you must get down-and-dirty in memory and look at a variable's value while the code runs. The Code::Block's debugger allows you to watch any variable in a program, showing you that variable's contents as the program runs. Listing 25-2 assists in this process:

LISTING 25-2: Where Variables Lurk

```
#include <stdio.h>
#include <stdlib.h>

int main()
{
    int x;
    int *px;

    px=&x;
    for(x=0;x<10;x++)
        printf("%d\n",*px);
    return 0;
}
```

Exercise 25-2: Create a new Code::Blocks project, `ex2502`, with debugging active. Copy the source code from Listing 25-2 into the `main.c` file. Build and run.

The code runs fine, but the debugger can help you see how it works. Further, you can witness how a pointer works in memory, which may help you understand the whole pointer thing. Regardless, it's time to debug! Follow these steps:

1. **Click the mouse to place it at the start of Line 6, where the integer variable `x` is declared.**
2. **Click the Run to Cursor button on the Debugging toolbar.**
3. **Click the Debugging Windows button.**

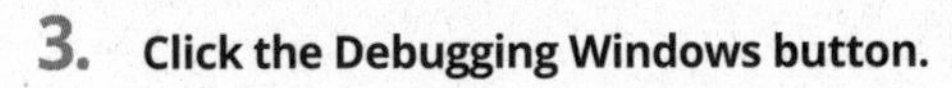

   This button is shown in the margin. It's really a menu, listing useful items.
4. **Choose the Watches command.**

   The Watches window appears, similar to the one shown in Figure 25-3. You see variables `x` and `px` listed under the Locals heading. You can add more variables to watch by typing them in.

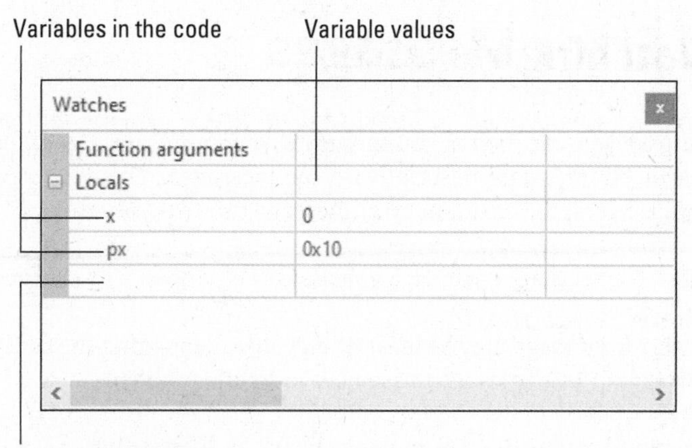

FIGURE 25-3: Monitoring variable values.

5. **Click the mouse in the first empty box in the Locals column in the Watches window.**

   Use Figure 25-3 as your guide.

6. **Type** *px **to examine the contents of the memory location stored in pointer** px**. Press the Enter key.**

   Because pointer px isn't assigned a value, an error message appears in the contents column. Until a variable is initialized, its contents are junk.

7. **Click the Next Line button on the Debugging toolbar until the cursor is on Line 10, the start of the *for* loop.**

   Because the value of pointer px is assigned to the memory location of variable x, pay heed to the Watches window. Instantly, you see a memory address appear by variable px, and you see the *px variable set equal to whatever value is assigned to variable x. The pointer has been initialized!

8. **Click the Next Line button repeatedly and observe the values in the Watches window.**

   As the *for* loop starts, it initializes the value of variable x. You see that value change in the Watches window, along with the value of *px. The value of px (the address of variable x) doesn't change.

9. **Click the Stop button when you're done.**

I find that examining variables in memory is the best way to see what's going on with your code. If the variables aren't popping the way they should, you need to check the statements manipulating those variables.

Improved Error Messages

One way you can better communicate your program's goof-ups to users is to present better, more descriptive error messages. Though too many details can confuse users, too scant an error message can frustrate them. For example:

```
Unable to allocate 128K char buffer 'input' at location 0xFE3958
```

This error message may be ideal when you're debugging the code, but a user either ignores it or "Googles it" to see how to fix the problem.

The opposite type of error message is just as frustrating:

```
Error 1202
```

For heaven's sake, don't use numbers as error messages! Even if you've provided documentation, no user appreciates it, especially when you can just as easily write

```
Not enough memory available
```

To help you craft better error messages, many C language functions — specifically, the file access functions — provide a consistent set of error values when a function fails. The error value is stored in the global variable `errno`, which your program can examine. Listing 25-3 provides sample source code.

LISTING 25-3: Checking the *errno* Value

```
#include <stdio.h>
#include <stdlib.h>
#include <errno.h>

int main()
{
    int e;

    e = rename("blorfus","fragus");
    if( e != 0 )
    {
        printf("Error! ");
        switch(errno)
```

```
        {
            case EPERM:
                puts("Operation not permitted");
                break;
            case ENOENT:
                puts("File not found");
                break;
            case EACCES:
                puts("Permission denied");
                break;
            case EROFS:
                puts("Read only file");
                break;
            case ENAMETOOLONG:
                puts("Filname is too long");
                break;
            default:
                puts("Too ugly to describe");
        }
    exit(1);
    }
    puts("File renamed");
    return 0;
}
```

To use the errno variable, your code must include the errno.h header file, as shown in Line 3 of Listing 25-3. That header file declares errno as an external variable. It also contains common error conditions as defined constants known to numerous C library functions.

The *rename()* function at Line 9 attempts to rename a file. I'm assuming that the file blorfus is unavailable, so the function is designed to generate an error. If so, it returns −1 and sets the global variable errno to represent the error condition that's encountered.

The *switch* structure at Line 13 plows through some errors that are possible when renaming a file. The defined constants represent these error codes, which are defined in the errno.h header file.

Exercise 25-3: Type the source code from Listing 25-3 into a new project. Build and run to witness an accurate error message.

You can refine the error messages further, if you like. I kept the messages short in Listing 25-3 so that the text wouldn't wrap in this book. For example, a better message than "Permission denied" is "The permission settings for 'blorfus' do not allow renaming. Consider resetting the file's permissions and trying again." This message is descriptive and informative, which is what users appreciate: It explains the problem and also offers a solution.

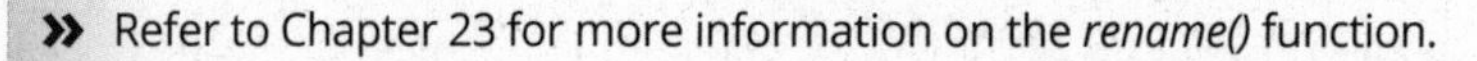

TIP

» Refer to Chapter 23 for more information on the *rename()* function.

» You can find relevant `errno` defined constants in a function's *man* page. Look there to see which defined constants to check and what they represent.

6 The Part of Tens

IN THIS PART . . .

See how to avoid ten common bugs and coding mistakes

Review ten helpful suggestions, reminders, and bits of advice

» == v. =

» Dangerous loop semicolons

» Commas in *for* loops

» Missing break in a switch structure

» Missing parentheses and curly brackets

» Pay heed to that warning

» Endless loops

» *scanf()* blunders

» Streaming input restrictions

Chapter 26
Ten Common Boo-Boos

The programming adventure has its pitfalls. Many of them are common; the same mistakes, over and over. Even after decades of coding, I still find myself doing stupid, silly things. Most often they're done in haste — usually, simple things that I'm not paying attention to. But isn't that the way of everything?

Conditional Foul-Ups

When you employ an *if* statement or initiate a *while* or *for* loop, the code makes a *comparison*. Properly expressing this comparison is an art form, especially when you try to do multiple things at once.

My advice is to first split up the code before you load everything into the parentheses. For example:

```
while((c=fgetc(dumpme)) != EOF)
```

The code on the preceding line works, but before you get there, try this:

```
c = 1;              /* initialize c */
while(c != EOF)
    c=fgetc(dumpme);
```

This improvement is more readable and less error prone.

The situation grows more hair when you use logical operators to combine conditions. I highly recommend that you limit the expression to two choices only:

```
if( a==b || a==c)
```

The statements belonging to *if* are executed when the value of variable a is equal to either the value of variable b or variable c. Simple. But what about this:

```
if( a==b || a==c && a==d)
```

Oops. Now the order of precedence takes over, and for the *if* statement to be true, a must be equal to b, or a must be equal to both c and d. The situation can also be improved by using parentheses:

```
if( a==b || (a==c && a==d))
```

REMEMBER

When you're unsure about the order of precedence, use parentheses.

== v. =

A single equal sign is the assignment operator:

```
a=65;
```

A double equal sign is used for comparison:

```
a==65
```

To get my brain to appreciate the difference, in my head I say "is equal to" when I type two equal signs. Despite this reminder, I still goof up, especially in a conditional statement.

WARNING

When you assign a value in a conditional statement, you generally create a TRUE condition unless the result of the assignment is zero:

```
if(here=there)
```

This *if* statement evaluates to the value of variable `there`. If it's 0, the *if* condition evaluates as FALSE; otherwise, it's true. Regardless, it's most likely not what you intended. Most compilers, such as *clang,* catch this mistake.

Dangerous Loop Semicolons

You can get into a rut when you're typing code, using the semicolon as a prefix before typing the Enter key to end a line. This practice has unintended consequences! It's especially toxic when you code loops:

```
for(x=0;x<10;x++);
```

This loop works, but its statements aren't repeated. When it's done, the value of variable `x` is `10`. That's it.

The same problem exists for a *while* loop:

```
while(c<255);
```

This loop may spin endlessly, depending on the value of variable `c`. If the value is greater than or equal to 255, the loop won't execute. Otherwise, it executes forever.

These semicolons are unintentional and unwanted. They're also perfectly legitimate; the compiler doesn't flag them as errors, though you may see a warning — which is quite helpful. For example, the suggested resolution is to place the semicolon on the next line:

```
while(putchar(*(ps++)))
    ;
```

This sole semicolon on a line by itself shouts to the compiler, as well as to any programmer reading the code, that the *while* loop is intentionally empty.

Commas in *for* Loops

The three items in a *for* loop's parentheses are separated by semicolons. Both those semicolons are required, and they are not commas. The compiler doesn't like this statement:

```
for(a=0,a<10,a++)
```

Because commas are allowed in a *for* statement, the compiler merely thinks that you've omitted the last two required items. In fact, the following is a legitimate *for* statement:

```
for(a=0,a<10,a++;;)
```

The preceding statement assigns the value 0 to variable a, generates a TRUE comparison (which is ignored), and increments the value of a to 1. Then, because the second and third items are empty, the loop repeats endlessly. (Well, unless a *break* statement belongs to the loop.)

Missing break in a switch Structure

It's perfectly legitimate to write a *switch* structure where the execution falls through from one case statement to the other:

```
switch(letter)
{
    case 'A':
    case 'E':
    case 'I':
    case 'O':
    case 'U':
        printf("Vowel");
        break;
    default:
        printf("Not a vowel");
}
```

In this example, the first five *case* conditions capture the same set of statements. When you forget the *break*, however, execution falls through with more tests and, eventually, the *default*. Unless this control is what you want, remember to add *break* statements as necessary.

Suppose that you have a *switch* structure that's several dozen lines high. One *case* condition has multiple rows of statements, so many that it scrolls up and off the screen. In such a setup, it's easy to forget the *break* as you concentrate instead on crafting the proper statements. I know — I've done it.

Another situation happens when you code a loop inside a *switch* structure and you use *break* to get out of that loop. This inner *break* escapes from the *for* loop only: A second *break* is required in order to get out of the *switch* structure.

TIP

Many editors, such as the one used in Code::Blocks, let you collapse and expand parts of your code. To make this feature work in a *switch* structure, you must enclose the *case* statements in curly brackets.

Missing Parentheses and Curly Brackets

Forgetting a parenthesis or two is one of the most common C coding mistakes. The compiler catches it, but usually the error isn't flagged until the end of the function.

For example, a missing parenthesis in the *main()* function causes the error to be flagged at the final line of the function. This warning is a good clue to a missing parenthesis or curly bracket, but it doesn't help you locate it.

Today's editors are good at matching up parentheses and brackets. For example, the Code::Blocks editor inserts both characters when you type the first one. This feature helps keep things organized. Other editors, such as *VIM*, highlight both sets of brackets when the cursor hovers over one. Use these hints as you type to ensure that things match up.

Another editor clue is that the formatting, text coloring, and indents screw up when you forget a parenthesis or bracket. The problem with recognizing this reminder is that the human brain automatically assumes that the editor has screwed up. So you need to train yourself to recognize improper indentation by the editor as a sign of a missing parenthesis or curly bracket.

Don't Ignore a Warning

When the compiler generates a warning, the program (or object code) is still created. This condition can be dangerous, especially when dealing with pointer errors. The problem is that warnings can be ignored; the code compiles anyway.

For example, you may be using *printf()* to display a value that you know is an *int*, but somehow the compiler insists that it's some other value. If so, you can type-cast the variable as an *int*. For example:

```
printf("%-14s %5ld %s",
    file->d_name,
    (long)filestat.st_size,
    ctime(&filestat.st_mtime));
```

In this example, the `filestate.st_size` variable is of the *off_t* data type. The *printf()* function lacks a conversion character for *off_t*, so it has typecast it to a *long int*. Similar typecasting can be done with other variable types for which *printf()* lacks a conversion character. But before you go nuts with this trick, check the man page for *printf()* to ensure that the specific data type lacks a conversion character.

TIP

- » A common warning happens when you try to display a *long int* value by using the `%d` placeholder. When that happens, just edit `%d` to `%ld`.
- » An "lvalue required" warning indicates that you've written a malformed equation. The *lvalue* is the left value, or the item on the left side of the equation. It must be present and be of the proper type so that the equation is properly handled.
- » The degree to which the compiler flags your code with warnings can be adjusted. Various *flags* are used to adjust the compiler's warning level. These flags are set in the Code::Blocks IDE by choosing the Project ⇨ Build Options command. The Compiler Flags tab in the Project Build Options dialog box lets you set and reset the various warnings.
- » The "turn on all warnings" option for a command-line C compiler is the `-Wall` switch. It looks like this:

  ```
  clang -Wall source.c
  ```

- » *Wall* stands for "warnings, all."

Endless Loops

There's got to be a way outta here, which is true for just about every loop. The exit condition must exist. In fact, I highly recommend that when you set out to code a loop, the *first thing* you code is the exit condition. As long as it works, you can move forward and write the rest of the joyous things that the loop does.

Unintentional endless loops do happen. I've run code many times, only to watch a blank screen for a few moments. Oops.

TIP

Console applications, such as the kind created throughout this book, are halted by pressing the Ctrl+C key combination in a terminal window. This trick may not always work, so you can try closing the window. You can also kill the task, which is a process that's handled differently by every operating system. For example, in a Unix operating system, you can open another terminal window and use the *kill* command to rub out a program run amok in the first terminal window.

scanf() Blunders

The *scanf()* function is a handy way to read specific information from standard input. It's not, however, ideal for all forms of input.

For example, *scanf()* doesn't understand when a user types something other than the format that's requested. Specifically, you cannot read in a full string of text. This issue is because *scanf()* discards any part of the string after the first whitespace character.

REMEMBER

Though the *fgets()* function is a great alternative for capturing text, keep in mind that it can capture the newline that ends a line of text. This character, `\n`, becomes part of the input string.

The other thing to keep in mind when using *scanf()* is that its second argument is an address, a pointer. For standard variable types — such as *int*, *float*, and *double* — you must prefix the variable name with the `&`, the address operator:

```
scanf("%d",&some_int);
```

The `&` prefix isn't necessary for reading in an array:

```
scanf("%s",first_name);
```

Individual array elements, however, aren't memory locations, and they still require the `&` prefix:

```
scanf("%c",&first_name[0]);
```

Pointer variables do not require the `&` prefix, which could result in unintended consequences.

Streaming Input Restrictions

The basic input and output functions in the C language aren't interactive. They work on streams, which are continuous flows of input or output, interrupted only by an end-of-file marker or, occasionally, the newline character.

When you plan to read only one character from input, be aware that the Enter key, pressed to process input, is still waiting to be read from the stream. A second input function, such as another *getchar()*, immediately fetches the Enter key press (the `\n` character). It does not wait, as an interactive input function would.

TIP

- If you desire interactive programs, get a library with interactive functions, such as the NCurses library. You can check out my book on Ncurses, available from Amazon.com in the kindle eBook and printed formats.
- The end-of-file marker is represented by the `EOF` constant, defined in the `stdio.h` header file.
- The newline character is represented by the `\n` escape sequence.

WARNING

- The newline character's ASCII value may differ from machine to machine, so always specify the escape sequence `\n` for the newline.

IN THIS CHAPTER

» Use creative names

» Write functions when needed

» Work on your code a little bit at a time

» Break apart larger projects into several modules

» Know what a pointer is

» Use whitespace before condensing

» Know when *if-else* becomes *switch-case*

» Remember to use assignment operators

» When you get stuck, read your code aloud

Chapter 27
Ten Reminders and Suggestions

It's difficult to narrow down the list of reminders and suggestions, especially for a topic as rich and diverse as programming. For example, I could suggest ways to fit in with other programmers, which movies to quote, which games to play, and even which foods to eat. A programming subculture exists — even today, though the emphasis on professional workplace attire has thankfully abated.

Beyond social suggestions, I do have a few things to remind you of — plus, some general-purpose C language recommendations. Believe it or not, every programmer has been through the same things you've experienced. It's good to hear advice from a grizzled programming veteran.

Maintain Good Posture

I'm certain that some authority figure somewhere in your early life drilled into you the importance of having proper posture. Ignore them at your own peril, especially when you're young and haven't yet gotten out of bed to say, "Ouch."

For many programmers, coding becomes an obsession. As an example, it's quite easy for me to sit and write code for many hours straight. Such a stationary position is hard on the body. So, every few minutes, take a break. If you can't manage that, schedule a break. Seriously: The next time you compile, stand up! Look outside! Walk around a bit!

While you're working, try as hard as you can to keep your shoulders back and your wrists elevated. Don't crook your neck when you look at the monitor. Don't hunch over the keyboard. Look out a window to change your focus.

REMEMBER

I might also add that it's pleasant to acknowledge others. True, it's easy to grunt or snarl at someone when you're in the midst of a project. Keep in mind that other humans may not appreciate the depth of thought and elation you feel when you code. If you can't be pleasant now, apologize later.

Use Creative Names

The best code I've seen reads like a human language. It's tough to make the entire source code read that way, but for small snippets, having appropriate variable and function names is a boon to writing clear code.

For example, the following expression is one of my favorites:

```
while(!done)
```

I read this statement as "while not done." It makes sense. Until the value of the `done` variable is TRUE, the loop spins. But somewhere inside the loop, when the exit condition is met, the value of `done` is set equal to TRUE and the loop stops. It's lucid.

It also helps to offer descriptive names to your functions. A name such as *setringervolume()* is great, but the name *set_ringer_volume()* is better. It also helps to consider the function in context. For example:

```
ch=read_next_character();
```

In the preceding line, the function *read_next_character()* needs no explanation — unless it doesn't actually return the next character.

Write a Function

Anytime you use code more than once, consider throwing it off into a function, even if the code is only one line long or appears in several spots and doesn't really seem function-worthy.

Suppose that you use the *fgets()* function to read a string, but then you follow *fgets()* with another function that removes the final newline character from the input buffer. Why not make both items their own function, something like *get_input()*?

Work on Your Code a Little Bit at a Time

A majority of the time you spend coding is to fix problems, to correct flaws in logic, or to fine-tune. When making such adjustments, avoid the temptation to make three or four changes at one time. Address issues one at a time.

The reason for my admonition is that it's tempting to hop around your code and work on several things at a time. For example: You need to fix the spacing in a *printf()* statement's output, adjust a loop, and set a new maximum value. Do those things one at a time!

When you attempt to do several things at a time, you can screw up. But which thing did you goof up? You have to go back and check everything, including the related statements and functions, to ensure that they work. During situations like these, you will seriously wish for a time machine. Instead, just work on your code a little bit at a time.

Break Apart Larger Projects into Several Modules

No one likes to scroll through 500 lines of code. Unless you're totally immersed in your project and can keep everything stored in your noggin, break out functions into modules.

I prefer to group related functions into similar files. I typically have an output file, an input file, an initialization file, and so on. Each file, or *module*, is compiled and linked separately to form the code. The benefits are that the files are smaller and if they compile and work, you no longer need to mess with them.

Know What a Pointer Is

A *pointer* is a variable that stores a memory location. It's not magic, and it shouldn't be confusing, as long as you keep the basic mantra in your head:

A pointer is a variable that stores a memory location.

A memory location stored in a pointer references another variable or a buffer (like an array). Therefore, the pointer must be initialized before it's used:

A pointer must be initialized before it's used.

When the pointer variable is prefixed by the * (asterisk) operator, it references the contents of the variable at the memory location. This duality is weird, of course, but it's highly useful, as demonstrated in Chapters 18 and 19.

TIP

- Declare a pointer variable by using the * prefix.
- Use the & operator to grab the address of any variable in C.
- Arrays are automatically referenced by their memory locations, so you can use an array name without the & prefix to grab its address.
- "Address" and "memory location" are the same thing.
- A great way to explore pointers is to use the Code::Blocks debugger; specifically, the Watches window. See Chapter 25.

Add Whitespace before Condensing

C programmers love to bunch up statements, cramming as many of them as they can into a single line. Even I am guilty of this pleasure, as shown by a few examples from this book, such as

```
while(putchar(*(sample++)))
```

Admit it: Such a construction looks cool. It makes it seem like you *really* know how to code C. But it can also be a source of woe.

My advice: Split out the code before you condense it. Make liberal use of whitespace, especially when you first write the code. For example, the line

```
if( c != '\0' )
```

is easier to read than the line

```
if(c!='\0')
```

After you write your code with whitespace — or use several statements to express something — you can condense, move out the spaces, or do whatever else you like.

REMEMBER

In C language source code, whitespace is for the benefit of human eyes. I admire programmers who prefer to use whitespace over cramming their code onto one line, despite how interesting it looks.

Know When *if-else* Becomes *switch-case*

I'm a big fan of the *if-else* decision tree, but I generally avoid stacking up multiple *if* statements. To me, it usually means that my programming logic is flawed. For example:

```
if(something)
    ;
else if(something_else)
    ;
else(finally)
    ;
```

This structure is okay, and it's often necessary to deal with a 3-part decision. But the following structure, which I've seen built by many budding C programmers, probably isn't the best way to code a decision tree:

```
if(something)
    ;
else if(something_else_1)
    ;
else if(something_else_2)
    ;
else if(something_else_3)
    ;
else if(something_else_4)
    ;
else(finally)
    ;
```

Generally speaking, anytime you have that many *else-if* statements, you probably need to employ the *switch-case* structure instead. In fact, my guess is that this example is probably what inspired the *switch-case* structure in the first place.

See Chapter 8 for more information on *switch-case*.

Remember Assignment Operators

Though it's nice to write readable code, one handy tool in the C language is an assignment operator. Even if you don't use one, you need to be able to recognize it.

The following equation is quite common in programming:

```
a = a + n;
```

In C, you can abbreviate this statement by using an assignment operator:

```
a += n;
```

REMEMBER

The operator goes before the equal sign. If it went afterward, it might change into a unary operator, which looks weird:

```
a =+ n;
```

So the value of variable a equals positive n? The compiler may buy that argument, but it's not what you intended.

Also, don't forget the increment and decrement operators, ++ and --, which are quite popular in loops.

When You Get Stuck, Read Your Code Out Loud

TIP

To help you track down that bug, start reading your code aloud. Pretend that a programmer friend is sitting right next to you. Explain what your code is doing and how it works. As you talk through your code, you'll find the problem. If you don't, have your imaginary friend ask you questions during your explanation.

Don't worry about going mental. You're a programmer. You're already mental.

As a bonus, talking through your code also helps you identify which portions need to have comments and what the comments should be. For example:

```
a++;      /* increment a */
```

In the preceding line, you see a terrible example of a comment. Duh. Of course, a is incremented. Here's a better version of that comment:

```
a++;      /* skip the next item to align output */
```

Don't just comment on what the code is doing — comment on *why*. Again, pretend that you're explaining your code to another programmer — or to future-you. Future-you will thank present-you for the effort.

7 Appendices

IN THIS PART . . .

Codes and keywords

Operators and data types

Escape sequences and conversion characters

Order of precedence

Appendix A

ASCII Codes

Decimal	Hex	Character	Comment
0	0x00	^@	Null, \0
1	0x01	^A	
2	0x02	^B	
3	0x03	^C	
4	0x04	^D	
5	0x05	^E	
6	0x06	^F	
7	0x07	^G	Bell, \a
8	0x08	^H	Backspace, \b
9	0x09	^I	Tab, \t
10	0x0A	^J	
11	0x0B	^K	Vertical tab, \v
12	0x0C	^L	Form feed, \f
13	0x0D	^M	Carriage return, \r
14	0x0E	^N	
15	0x0F	^O	
16	0x10	^P	
17	0x11	^Q	
18	0x12	^R	
19	0x13	^S	
20	0x14	^T	
21	0x15	^U	

Decimal	Hex	Character	Comment
22	0x16	^V	
23	0x17	^W	
24	0x18	^X	
25	0x19	^Y	
26	0x1A	^Z	
27	0x1B	^[	Escape
28	0x1C	^\	
29	0x1D	^]	
30	0x1E	^^	
31	0x1F	^_	
32	0x20		Space, start of visible characters
33	0x21	!	Exclamation point
34	0x22	"	Double quote
35	0x23	#	Pound, hash
36	0x24	$	Dollar sign
37	0x25	%	Percent sign
38	0x26	&	Ampersand
39	0x27	'	Apostrophe
40	0x28	(	Left parenthesis
41	0x29	)	Right parenthesis
42	0x2A	*	Asterisk
43	0x2B	+	Plus
44	0x2C	,	Comma
45	0x2D	-	Hyphen, minus
46	0x2E	.	Period
47	0x2F	/	Slash
48	0x30	0	Numbers

Decimal	Hex	Character	Comment
49	0x31	1	
50	0x32	2	
51	0x33	3	
52	0x34	4	
53	0x35	5	
54	0x36	6	
55	0x37	7	
56	0x38	8	
57	0x39	9	
58	0x3A	:	Colon
59	0x3B	;	Semicolon
60	0x3C	<	Less than, left angle bracket
61	0x3D	=	Equals
62	0x3E	>	Greater than, right angle bracket
63	0x3F	?	Question mark
64	0x40	@	At sign
65	0x41	A	Uppercase alphabet
66	0x42	B	
67	0x43	C	
68	0x44	D	
69	0x45	E	
70	0x46	F	
71	0x47	G	
72	0x48	H	
73	0x49	I	
74	0x4A	J	
75	0x4B	K	

Decimal	Hex	Character	Comment
76	0x4C	L	
77	0x4D	M	
78	0x4E	N	
79	0x4F	O	
80	0x50	P	
81	0x51	Q	
82	0x52	R	
83	0x53	S	
84	0x54	T	
85	0x55	U	
86	0x56	V	
87	0x57	W	
88	0x58	X	
89	0x59	Y	
90	0x5A	Z	
91	0x5B	[	Left square bracket
92	0x5C	\	Backslash
93	0x5D	]	Right square bracket
94	0x5E	^	Caret
95	0x5F	_	Underscore
96	0x60	`	Back tick, accent grave
97	0x61	a	Lowercase alphabet
98	0x62	b	
99	0x63	c	
100	0x64	d	
101	0x65	e	

Decimal	Hex	Character	Comment
102	0x66	f	
103	0x67	g	
104	0x68	h	
105	0x69	i	
106	0x6A	j	
107	0x6B	k	
108	0x6C	l	
109	0x6D	m	
110	0x6E	n	
111	0x6F	o	
112	0x70	p	
113	0x71	q	
114	0x72	r	
115	0x73	s	
116	0x74	t	
117	0x75	u	
118	0x76	v	
119	0x77	w	
120	0x78	x	
121	0x79	y	
122	0x7A	z	
123	0x7B	{	Left brace, left curly bracket
124	0x7C	|	Vertical bar
125	0x7D	}	Right brace, right curly bracket
126	0x7E	~	Tilde
127	0x7F		Delete

- ASCII 0 through ASCII 31 represent control code values. These characters are accessed by pressing the Ctrl key on the keyboard and typing the corresponding symbol or letter key.
- Code 32 is the code for the space character.
- Code 127 is the Delete character, which is different from Code 8, Backspace. The reason is that Code 8 is defined as *nondestructive,* which means that it only moves the cursor back a space.
- Many of the control codes manipulate text on the screen, such as Ctrl+I for the Tab key.
- A keen eye can spot three repetitions in the ASCII code lists. Look at codes 0 through 26 and then 64 through 90. Also look at codes 97 through 122.
- The difference between uppercase and lowercase characters in the table is 32, a computer holy number. The hexadecimal difference is 0x20. Therefore, by using simple math, or *bitwise logic,* you can convert between upper- and lowercase.
- The digits 0 through 9 are equal to the values `0` through `9` when you subtract 48 (0x30) from the ASCII code values. Likewise, to convert values `0` through `9` into their corresponding ASCII characters, add 48 or 0x30.
- Any ASCII character can be represented as an escape sequence. Follow the backslash with the character's code value, as in \33 for the exclamation point (`!`) character. The hexadecimal value can also be used, as in \x68 for the little *H.*

Appendix B

Keywords

C Language Keywords, C17 Standard

_Alignas	break	float	signed
_Alignof	case	for	sizeof
_Atomic	char	goto	Static
_Bool	const	if	Struct
_Complex	continue	inline	Switch
_Generic	default	int	Typedef
_Imaginary	do	long	Union
_Noreturn	double	register	unsigned
_Static_assert	else	restrict	void
_Thread_local	enum	return	Volatile
auto	extern	short	While

Deprecated C Language Keywords, No Longer Standard

asm	entry	Fortran

C++ Language Keywords

Asm	dynamic_cast	new	requires	typeid
Bool	export	not	static_cast	typename
Catch	False	operator	template	using
Class	friend	private	this	virtual
Concept	inline	protected	throw	xor
const_cast	mutable	public	true	
Delete	namespace	reinterpret_cast	try	

- The C17 standard is the current C language standard as this book goes to press. The standard was established in 2017.
- You don't need to memorize the C++ keywords, and not all of them are listed in this appendix. Still, it's important to be aware of them. I strongly recommend that you not use any of them as function names or variable names in your code.
- The most frequent C++ reserved word that C programmers tend to use is *new*. Just don't use it; use something else, like *new_item* or *newSomething* instead.
- The *bool* keyword in C++ is effectively the same thing as the *_Bool* keyword in C.

Appendix C

Operators

Also see Appendix G for the order of precedence.

Operator	Type	Function
+	Math	Addition
–	Math	Subtraction
*	Math	Multiplication
/	Math	Division
%	Math	Modulo
++	Math	Increment
--	Math	Decrement
+	Math	Unary plus
-	Math	Unary minus
=	Assignment	Assigns a value to a variable
+=	Assignment	Addition
-=	Assignment	Subtraction
*=	Assignment	Multiplication
/=	Assignment	Division
%=	Assignment	Modulo
!=	Comparison	Not equal
<	Comparison	Less than
<=	Comparison	Less than or equal to
==	Comparison	Is equal to
>	Comparison	Greater than
>=	Comparison	Greater than or equal to

<table>
<tr><th>Operator</th><th>Type</th><th>Function</th></tr>
<tr><td>?:</td><td>Ternary</td><td>Either-or decision</td></tr>
<tr><td>.</td><td>Structure</td><td>Member</td></tr>
<tr><td>-></td><td>Structure</td><td>Member of a pointer structure</td></tr>
<tr><td>&&</td><td>Logical</td><td>Both comparisons are true</td></tr>
<tr><td>||</td><td>Logical</td><td>Either comparison is true</td></tr>
<tr><td>!</td><td>Logical</td><td>The item is false</td></tr>
<tr><td>&</td><td>Bitwise</td><td>Mask bits</td></tr>
<tr><td>|</td><td>Bitwise</td><td>Set bits</td></tr>
<tr><td>^</td><td>Bitwise</td><td>Exclusive or (XOR)</td></tr>
<tr><td><<</td><td>Bitwise</td><td>Shift bits to the left</td></tr>
<tr><td>>></td><td>Bitwise</td><td>Shift bits to the right</td></tr>
<tr><td>~</td><td>Unary</td><td>One’s complement</td></tr>
<tr><td>!</td><td>Unary</td><td>NOT</td></tr>
<tr><td>*</td><td>Unary</td><td>Pointer (peeker)</td></tr>
</table>

Appendix D

Data Types

Standard Data Types

Type	Value Range	Conversion Character
void	None	None
_Bool	0 to 1	%d
char	–128 to 127	%c
unsigned char	0 to 255	%u
short int	–32,768 to 32,767	%d
unsigned short int	0 to 65,535	%u
int	–2,147,483,648 to 2,147,483,647	%d
unsigned int	0 to 4,294,967,295	%u
long int	–2,147,483,648 to 2,147,483,647	%ld
unsigned long int	0 to 4,294,967,295	%lu
long long	-9,223,372,036,854,775,808 to 9,223,372,036,854,775,807	%lld
unsigned long long	0 to 18,446,744,073,709,551,615	%llu
float	1.2×10^{-38} to 3.4×10^{38}	%e, %f, %g
Double	2.3×10^{-308} to 1.7×10^{308}	%e, %f, %g
long double	3.4×10^{-4932} to 1.1×10^{4932}	%e, %f, %g

The `%i` placeholder generates integer output, the same as `%d`. (Think `%i` for integer, but `%d` for decimal.)

- The `%x` placeholder outputs integer values in hexadecimal. When `%X` is specified, numbers A through F are output in uppercase.
- The `%o` placeholder outputs integer values in octal.

- Some overlap exists between the size of an *int* and the size of a *long*. The ranges depend on how these data types are implemented on a given system.
- The `limits.h` header file lists the sizes of data types.
- The *size_t* defined data type represents bytes in memory. This *typedef* value is nominally a *char*, but *size_t* is used instead of *char* (or *unsigned char*). The placeholder to represent this value is `%z`, though `%zu` (unsigned) and `%zd` (decimal output) are also used.

Appendix E

Escape Sequences

Characters	What It Represents or Displays
\a	Bell ("beep!")
\b	Backspace, non-erasing
\f	Form feed or clear the screen
\n	Newline
\r	Carriage return
\t	Tab
\v	Vertical tab
\\	Backslash character
\?	Question mark
\'	Single quote
\"	Double quote
\x*nn*	Hexadecimal character code *nn*
\o*nn*	Octal character code *nn*
nn	Octal character code *nn*

Appendix F

Conversion Characters

Conversion Character	What It Displays
%%	Percent character (%)
%c	Single character (char)
%d	Integer value (short, int)
%e	Floating-point value in scientific notation using a little E (float, double)
%E	Floating-point value in scientific notation using a big E (float, double)
%f	Floating-point value in decimal notation (float, double)
%g	Substitution of `%f` or `%e`, whichever is shorter (float, double)
%G	Substitution of `%f` or `%E`, whichever is shorter (float, double)
%i	Integer value (short, int)
%ld	Long integer value (long int)
%o	Unsigned octal value; no leading zero
%p	Memory location in hexadecimal (*pointer)
%s	String (char *)
%u	Unsigned integer (unsigned short, unsigned int, unsigned long)
%x	Unsigned hexadecimal value, lowercase (short, int, long)
%X	Unsigned hexadecimal value, capital letters (short, int, long)

Conversion-character formatting

The options available for conversion characters in C are extensive. The *printf()* man page lists many of them, with some requiring a bit of experimentation to get them correct. Generally speaking, here's the format for the typical conversion character:

```
%-pw.dn
```

Only the first and last characters are required: `%` is the percent sign that prefixes all conversion characters, and *n* is the conversion character(s).

- The minus sign; works with the *w* option to right-justify output.

p The padding character, which is either zero or a space, when the *w* option is used. The padding character is normally a space, in which case the *p* need not be specified. When *p* is 0, however, the value is padded on the left with zeroes to match the width set by the *w* option.

w The width option; sets the minimum number of positions in which the information is displayed. Output is right-justified unless the – prefix is used. Spaces are padded to the left, unless the *p* value specifies the character 0 (a zero).

.d The dot, followed by a value, *d*, that describes how many digits to display after the decimal in a floating-point value. If *d* isn't specified, only the whole-number portion of the value appears.

n A conversion character, as shown in the table in this appendix. Or it can be the percent sign (%), in which case a `%` appears in the output.

Appendix G

Order of Precedence

- The *order of precedence* goes as shown in Table G-1 and is overridden by using parentheses. C always executes the portion of an equation in parentheses before anything else.
- Incremented or decremented variables as lvalues (assigned to another value) operate left-to-right. So, `++var` increments before its value is assigned; `var++` increments its value after it's assigned. (See Table G-2.)
- Associativity for the assignment operators moves right-to-left. For example, the operation on the right side of += happens first.
- The order of precedence may also be referred to as the *order of operations.*

TABLE G-1 **Standard Operator Precedence**

Operator(s)	Category	Description
() [] -> . (period)	Expression	Function arguments, arrays, pointer members
! ~ -- + * & ++ --	Unary	Logical not, one's complement, positive, negative, pointer, address-of, increment, decrement; operations right-to-left
* / %	Math	Multiplication, division, modulo
+ -	Math	Addition, subtraction
<< >>	Binary	Shift left, shift right
< > <= >=	Comparison	Less than, greater than, less than or equal to, greater than or equal to
== !=	Comparison	Is equal to, not equal to
&	Binary	And
^	Binary	Exclusive or (XOR)
|	Binary	Or

(continued)

TABLE G-1 ***(continued)***

<table>
<tr><th>Operator(s)</th><th>Category</th><th>Description</th></tr>
<tr><td>&&</td><td>Logical</td><td>And</td></tr>
<tr><td>||</td><td>Logical</td><td>Or</td></tr>
<tr><td>?:</td><td>Comparison</td><td>Ternary operator, associativity goes right-to-left</td></tr>
<tr><td>=</td><td>Assignment</td><td>Variable assignment operator, including += and *= and all assignment operators</td></tr>
<tr><td>, (comma)</td><td>(None)</td><td>Separates items in a for statement; precedence is left-to-right</td></tr>
</table>

TABLE G-2 **Pointers and Precedence**

Expression	Address p	Value *p
p	Yes	No
*p	No	Yes
*p++	Incremented after value is read	Unchanged
*(p++)	Incremented after value is read	Unchanged
(*p)++	Unchanged	Incremented after it's read
*++p	Incremented before value is read	Unchanged
*(++p)	Incremented before value is read	Unchanged
++*p	Unchanged	Incremented before it's read
++(*p)	Unchanged	Incremented before it's read

Index

Symbols

A

B

C

D

E

F

J

K

L

M

N

O

P

Q

R

T

U

V

W

X

Z

About the Author

Dan Gookin has been writing about technology for nearly three decades. He combines his love of writing with his gizmo fascination to create books that are informative, entertaining, and not boring. Having written over 160 titles with 12 million copies in print translated into over 30 languages, Dan can attest that his method of crafting computer tomes seems to work.

Perhaps his most famous title is the original *DOS For Dummies*, published in 1991. It became the world's fastest-selling computer book, at one time moving more copies per week than the *New York Times* number-one bestseller (though, as a reference, it could not be listed on the Times' Best Sellers list). That book spawned the entire line of *For Dummies* books, which remains a publishing phenomenon to this day.

Dan's most popular titles include *PCs For Dummies, Laptops For Dummies,* and *Microsoft Word For Dummies.* He also maintains the vast and helpful website `www.wambooli.com`.

Dan holds a degree in Communications/Visual Arts from the University of California, San Diego. He lives in the Pacific Northwest, where he enjoys spending time annoying people who deserve it.

Publisher's Acknowledgments

Acquisitions Editor: Katie Mohr

Senior Project Editor: Paul Levesque

Copy Editor: Becky Whitney

Production Editor: Tamilmani Varadharaj

Cover Image: © kasezo/Getty Images